Mapping Biological Systems to Network Systems

Heena Rathore

Mapping Biological Systems to Network Systems

 Springer

Heena Rathore
Indian Institute of Technology
Jodhpur
India

ISBN 978-3-319-29780-4   ISBN 978-3-319-29782-8   (eBook)
DOI 10.1007/978-3-319-29782-8

Library of Congress Control Number: 2016930677

Printed on acid-free paper

This Springer imprint is published by SpringerNature
The registered company is Springer International Publishing AG Switzerland

Contents

Chapter 1
Introduction: Bio-inspired Systems

Abstract Nature is the physical world around us in particular and life in general. It has immense matter to explore, analyze, and investigate. For instance, plants perform photosynthesis, bees search for nectar, birds fly in a synchronized way; the sun rises and sets in a specific way. On observing nature keenly, it can be inferred that it is perfect, divine, structured, and mannered. It is because of the intrinsic appealing characteristics of biological systems that nowadays many researchers are engaged in producing novel design paradigms to address the challenges in current network systems based on biological systems. Biologically inspired approaches seem promising when high level of robustness and adaptability is required. The chapter begins by exploring why biology and computer network research are such a natural match. This is followed by presenting a broad overview of biologically inspired research in network systems. It is classified by the biological field that inspired each topic and by the area of networking in which that topic lies. Each case elucidates how biological concepts have been most successfully applied in various domains.

1.1 Biological Systems

Bio-inspired systems are those systems where biology plays a vital part to solve problems in another domain. Some of the inherent characteristics of biological systems are listed below:

(1) Biological systems are flexible by nature and capable of adapting to dynamics of environmental conditions. They are able to resist the varied surroundings and atmosphere.
(2) They have a proven capacity to heal, remain strong, and be flexible to the failures caused by many natural and man-made factors.
(3) They are able to perform and accomplish very intricate tasks using a limited set of basic rules and fundamentals.
(4) When presented with novel situations, they are simple, convenient, and efficient in learning, resolving, and regenerating them.

© Springer International Publishing Switzerland 2016
H. Rathore, *Mapping Biological Systems to Network Systems*,
DOI 10.1007/978-3-319-29782-8_1

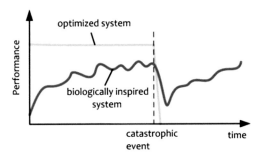

Fig. 1.1 Performance level: optimized system versus biologically inspired system (*Source* Leibnitz et al. 2007)

(5) They have effective expert management of accountable resources and coordination among the various components of their system in such a way that they can be called global intelligent systems.
(6) Since different components coordinate with each other, these systems can be called stable systems having efficient equilibrium.
(7) They are self-organized by nature and renewable in context.
(8) They can survive under extreme harsh conditions and adversities because of their inherent resistance despite having moderate resources.

One of the incredibly diverse characteristics of biological systems is that they are robust. They may exhibit inferior performance compared to optimized systems but have higher resilience toward critical errors as shown in Fig. 1.1 (Leibnitz et al. 2007).

Bio-inspired networking refers to fields where biology has inspired computer networks. An excellent survey of bio-inspired networks has been made by Meisel and others (2010) and Dressler and Akan (2010). They cover topics such as firefly synchronization, activator and inhibitor systems, swarm intelligence, artificial immune systems, epidemic spreading, and cellular signaling networks. These topics are covered in the following subsection.

The study of bio-inspired network systems explores the opportunities from biological systems by mimicking the laws and dynamics governing them and exhibiting a parallelism between the two fields of biology and networks. There are numerous topics in computer networks that can be studied from a biological perspective.

1.2 Network Systems

A computer network is a collection of connected computers that allow transmission of information between them. Computer networks have evolved from static systems to highly autonomous and dynamic networks. The evolving nature of network system poses numerous challenges that need to be resolved. Some of these challenges are:

- *Connectivity and scalability*: Size of the network is a major issue because of its high density and interconnectivity. Since the system is open, any number of nodes can be added onto it. Hence, it should be scalable in a way that it can acquire large-scale networking without hampering the normal functionality of the system.
- *Dynamism of the system*: Early communication systems, with a single transmitter, channel, and receiver, were static in nature. Static networks were stable because they were not prone to varying dynamics of the system. However, dynamic networks were prone to it in terms of behavior, traffic, bandwidth, channel, and network conditions.
- *Network resources*: They should be effectively used and managed to make the system cost-effective.
- *Need for infrastructureless and autonomous operation*.
- They must have the capability to *self-organize, self-evolve,* and *survive under dynamic situations*.

The aforementioned issues require a dedicated solution to resolve them. Biologically inspired approaches seem promising since they have the capability to self organize, self-survive, self-adapt to varying environmental conditions and by looking at the analogies between networks and biology, biological inspired solutions can be the best elucidation.

1.3   Mapping Biological Systems to Network Systems

In order to develop next-generation bio-inspired systems, it is essential to understand nature, which requires close collaboration with biologists. Consequently, the probable biological solutions can be applied to the respective field of computer networks and provide close interaction between the two components. The process is a four-step MOLE model with the following sequence:

- *Observe* natural behavior in parallel to desired network behavior.
- *Explore* basic biological behavior which includes components, process, and model.
- *Look* for specific behavior that optimally maps to networks.
- *Map* that behavior into a mathematical model for new implementations.

This approach can help in developing novel and intriguing results in the domain of biologically inspired systems. Networking researchers can develop better biologically inspired techniques with the help of system biologists, while improving their understanding of the biological networks. Nevertheless, such collaborations enhance the knowledge and understanding of complex and dynamic network systems.

1.4   Motivation

The study of bio-inspired network systems explores the opportunities from biological systems by mimicking the laws and dynamics governing them and
exhibiting a parallelism between the two fields of biology and networks. There are
numerous topics in computer networks that can be studied from a biological perspective as shown in Figs. 1.2 and 1.3. Figure 1.2 illustrates network research
inspired by biology for some of the topics, organized by the area of inspiration.
Similarly, Fig. 1.3 deals with an alternative view of network research inspired by
biology, organized by the area of application.

There are many issues in networks such as routing, admission control, and
quality of service which can be looked upon from the biological perspective. A few
bio-inspired systems, namely firefly synchronization, activator and inhibitor systems, swarm intelligence, artificial immune systems, epidemic spreading, and cellular signaling networks, are described below.

(1) *Swarm intelligence*: Collectively, social insects can perform various complex
    tasks such as food gathering, motion paths, etc. Swarm intelligence is a
    technique that takes inspiration from social insects such as ants, bees, and birds
    which perform group activities. The application of swarm intelligence in
    networks can be seen in the area of ad hoc routing, video streaming, Huffman
    coding, etc.

    a. Areas such as social insect routing, epidemic routing, and physiological
       networks provide extensive insights into how routing of information is
       handled in nature. Ant colony optimization (ACO) is a metaheuristic
       algorithm to find the shortest path in large graphs. Ants have an inbuilt

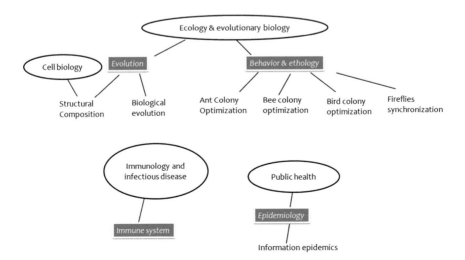

Fig. 1.2 Domains of biology inspired the research in networks

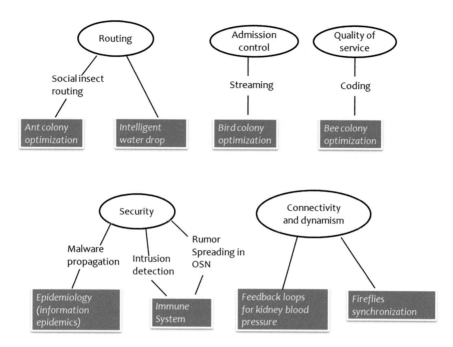

Fig. 1.3 Alternative view: domains of research in networks inspired from biology

capability to converge from their nest to the food source in the shortest possible way. This study can serve as a source of inspiration for routing messages between nodes using the shortest plausible distance (Blum 2005). This information can also be used for tackling social network problems (Ahmad and Srivastava 2008; Mandala et al. 2013).

b. When birds collectively flock together they exhibit a coordinated beautiful pattern. They follow a behavioral model in which each agent/bird follows separation, cohesion, and alignment with each other. Bird colonies were used for video streaming for improving quality of service (Baguda et al. 2012). The system performed better under bad channel conditions due to its simplicity and fast convergence property.

c. Bee colonies perform the forage selection, which is the act of looking or searching for food through a beautiful waggle dance. Interestingly, bee colonies were used for the generation of Huffman Codes (Tatar and Holban 2012). The algorithm inspired from bees performed better in terms of lookup time, which is significantly better than the traditional structural approach.

(2) *Firefly synchronization*: The phenomenon, where thousands of male fireflies of a certain species congregate in trees and flash in synchrony, is observed at night in certain parts of the world. Fireflies can simply be abstracted as oscillators that emit a pulse of light periodically. These types of oscillators are

referred to as pulse coupled oscillators and are used to study biological systems such as neurons and earthquakes. Applications in the field of networking include robust and fully distributed clock synchronization (Mirollo and Strogatz 1990). Fireflies can also be linked with the data gathering aspect of wireless sensor networks (Taniguchi et al. 2006).

(3) *Activator and inhibitor systems*: These are those that relate to reaction and diffusion mechanisms. An activator and inhibitor system consists of two substances that act on each other. The activator stimulates its own production via auto-catalysis as well as the production of the inhibitor. The inhibitor, in turn, represses the production of the activator. In addition, the inhibitor diffuses more rapidly than the activator such that the patterns of activator and inhibitor concentrations arise. These systems can be used for the self-organization of autonomous systems, distributed coordination, and continuous adaptation of system parameters in a highly dynamic environment (Henderson et al. 2004; Dressler 2008).

(4) *Artificial immune system*: Intrusion detection and propagation of viruses, i.e., areas related to security, can be studied from an immunological perspective (Elsadig and Abdullah 2008a, b; Boukerche et al. 2007). Here again, there are interesting natural phenomena, which can serve as inspiration. The human body has the ability to build immunity against many viruses. In addition, it also has the ability to fight back against a new virus. This can be used as inspiration to develop self-healing and self-adapting systems to address the issues in security (Wang et al. 2006; Wang and Suda 2001).

(5) *Epidemic spreading*: The study of epidemic disease has always been an area of research where biological issues are mixed with social ones. When epidemic disease is talked about, it is about contagious diseases like influenza, measles, and sexually transmitted diseases. Epidemics can either pass explosively through a population (Goffman and Newill 1964), or persist over a longer time period at low levels. They can experience unexpected flare-ups or even wavelike cyclic patterns of increasing and decreasing prevalence. This field can be directly linked to content distribution in computer networks (e.g., in Delay Tolerant Networks), overlay networks, and analysis of worm and virus spreading on the Internet. It can also be used in studying rumor spreading when it comes to online social networks (Doerr et al. 2012; David et al. 2005).

(6) *Cellular signaling networks*: These networks comprise of highly connected modules that can be utilized to manage multiple functions in a context-dependent approach. This field directly links with coordination and control in massively distributed systems and programming of network-centric operating sensor and actor networks (Tschudin 2003; Pawson 1995).

Biologically inspired approaches for resolving computational problems have recently gained importance in computing due to the need for flexible and adaptable ways for solving networking problems. Robotics is one such field which has gained importance and is rapidly growing (Beer et al. 1997; Cruse et al. 1998). Recently, honeybees have also been used as a model for the study of visually guided flight and navigation in robots (Srinivasan 2011). Consensus-based schemes are also being used to sense the surrounding environment (Yu et al. 2010). Nanonetwork is the technology used to create devices that are as small as the size of a human cell. This technology closely resembles biological cell communication and has recently shown some advances (Akyildiz et al. 2011). A review of the future prospects of bio-inspired networking is given by Nakano (2011).

Biological systems have inherent capabilities of evolving, self-organizing, self-repairing, and flourishing with time. These characteristics make biologically inspired approaches seem propitious. Bio-inspired systems refer to take the opportunities from the biology domain and map with the problems faced in the network domain. Since the issues pertaining to network domain require a robust solution, these approaches are encouraging. Although research efforts have been made with regard to various approaches, there are still some challenges that need to be addressed. There are various factors that influence a network design such as scalability, production cost, operating environment, hardware constraints, transmission media, power consumption, data processing, communicating, etc. All these factors have to be taken into account while designing a better and efficient network. Network performance mainly depends on three factors:

(1)   Quality of Service: Measures how well a network performs a task.
(2)   Fault Tolerance: Handles node failures without letting QoS fall below a certain level.
(3)   Lifetime of entire system: Time during which network QoS is above a certain level.

Novel design paradigm is required to challenge and face the issues in network systems. For this the prominent fit is biologically inspired approaches.

1.5   Organization of Book

The book primarily provides insight into bio-inspired systems. The outline for the book is described below:

Chapter 1: Introduction: Bio-inspired Systems

It focuses on mapping between the two systems and how the two systems are related to each other. It emphasizes some of the major problems in network systems and how biology has played a significant role in solving them.

Chapter 2: Computer Networks

The chapter focuses on network design, topology, challenges, and usage of network in the present scenario. It explains how the Open Systems Interconnect (OSI) model works to simplify networking and provides details of various networking topologies. Furthermore, issues and challenges in networking domain described in the chapter make them evident to have self-adaptive and self-healing solutions.

Chapter 3: Inceptive Findings

The chapter explores the analogies that can be derived structurally and operationally from biological systems. These findings impart the basic framework that can be derived from biological systems and how they can be integrated into network systems.

Chapter 4: Swarm Intelligence and Social Insects

The chapter explores analogies that can be derived from swarm intelligence and social insects. It explicates social insects such as ants, bees, birds, fireflies, and their mapping toward implementing solutions for network systems.

Chapter 5: Immunology and Immune System

The chapter delivers basic insight into biological immune systems which have intelligent capabilities of detecting foreign bodies that attack the human body. Later, it discusses the ways in which the immune system can be adopted in solving various problems.

Chapter 6: Information Epidemics and Social Networking

The chapter explores the domain of epidemic spreading. This chapter illustrates some of the novel work that can be adapted in solving issues in social networks such as rumor spreading, providing trust in the network, etc.

Chapter 7: Artificial Neural Network

The chapter investigates the domain of artificial neural network and the way they are utilized in the networking domain in resolving issues such as pattern recognition, clustering, optimization, etc.

Chapter 8: Genetic Algorithms

Genes are an important constituent of human bodies. The way in which nature and the human body evolve with time is beyond comparison. The present chapter provides insights into genetic algorithms and evolutionary computation. Bio-inspired techniques in this domain have shown much advancement which can be utilized in the networking environment.

Chapter 9: Bio-inspired Software Defined Networking

Software defined networking (SDN) has provided a breakthrough in network transformation. This chapter discusses about SDN and how biologically inspired techniques have helped in overcoming some of the issues.

Chapter 10: Case Study: A Review of Security Challenges, Attacks, Trust and Reputation Models in Wireless Sensor Networks

The chapter begins by explaining the security issues and challenges in WSN. It discusses the goals, threat models, and attacks followed by the security measures that can be implemented in the detection of attacks. Here, various types of trust and reputation models are reviewed. The intent of this chapter is to investigate the security-related issues and challenges in wireless sensor networks and the methodologies used to overcome them. The chapter focuses on reducing the effect of fraudulent nodes by the immune-inspired model. It gives basic insight into biological immune systems that have intelligent capabilities of detecting foreign bodies that attack the human body. Later, it discusses the methodologies for attaining the similar types of perspicacious nature that can be adapted in removal of fraudulent nodes in WSN. The chapter provides results that demonstrate the accuracy and robustness of the proposed model. It also provides a method for detection of the most reputable path using intelligent water drops.

Chapter 11: Bio-inspired Approaches in Various Engineering Domain

The chapter discusses various bio-inspired approaches in different engineering domains such as energy, agriculture, aerospace, electrical, etc. The context of providing such details is to show how bio-inspired approaches are not only dependent on computer networks but have also shown tremendous growth in other engineering aspects.

References

Ahmad, M. A., & Srivastava, J. (2008). An ant colony optimization approach to expert identification in social networks. In *Proceedings Social Computing, Behavioral Modeling and Prediction* (pp. 120–128).

Akyildiz, I. F., Jornet, J. M., & Pierobon, M. (2011). Nanonetworks: A new frontier in communications. *Communications of the ACM, 54*(11), 84–89.

Baguda, Y. S., Fisal, N., Rashid, R. A., Yusof, S. K., Syed, S. H., & Shuaibu, D. S. (2012). Biologically-inspired optimal video streaming over unpredictable wireless channel. *International Journal of Future Generation Communication and Networking, 5*(1), 15–28.

Beer, R. D., Quinn, R. D., Chiel, H. J., & Ritzmann, R. E. (1997). Biologically-inspired approaches to robotics. *Communications of the ACM, 4*(3), 30–38.

Blum, C. (2005). Ant colony optimization: Introduction and recent trends. *Physics of Life reviews Journal, 2*, 353–373.

Boukerche, A., Machado, R. B., Juca, K. R. L., Sobral, J. B. M., & Notare, M. S. (2007). An agent based and biological inspired real-time intrusion detection and security model for computer network operations. *Journal Computer Communications, 30*(13), 2649–2660.

Cruse, H., Kindermann, T., Schumm, M., Dean, J., & Schmitz, J. (1998). Walknet—a biologically inspired network to control six-legged walking. *Neural networks, 11*(7), 1435–1447.

David, K., Kleinberg, J., & Tardos, E. (2005). Influential nodes in a diffusion model for social networks. In *Proceedings of Automata, languages and programming* (pp. 1127–1138).

Doerr, B., Fouz, M., & Friedrich, T. (2012). Why rumors spread fast in social networks. *Magazine Communications of the ACM, 55*, 1–10.

Dressler, F. (2008). Self-organized event detection in sensor networks using bio-inspired promoters and inhibitors. In *Proceedings of International Conference on Bio-Inspired Models of Network, Information and Computing Systems* (Vol. 3, pp. 1–8).

Dressler, F., & Akan, O. B. (2010). A survey on bio-inspired networking. *Computer Networks: The International Journal of Computer and Telecommunications Networking, 54*(6), 881–900.

Elsadig, M., & Abdullah, A. (2008a). Biological inspired intrusion prevention and self-healing system for network security based on danger theory. *International Journal of Video and Image Processing and Network Security, 9*(9).

Elsadig, M., & Abdullah, A. (2008b). Biological inspired approach in parallel immunology system for network security. In *Proceedings on International Symposium on Information Technology* (Vol. 1, pp. 1–7).

Goffman, W., & Newill, V. (1964). Generalization of epidemic theory: An application to the transmission of ideas. *Nature, 11*(8), 204–225.

Henderson, T. C., Venkataraman R., & Choikim G. (2004). Reaction-diffusion patterns in smart sensor networks. In *Proceedings of IEEE Conference on Robotics and Automation* (Vol. 1, pp. 654–658).

Leibnitz, K., Wakamiya, N., & Murata, M. (2007). Biologically inspired networking. In *Cognitive Networks: Towards Self-Aware Network* (ch-1, pp. 1–21). Book-Wiley Interscience.

Mandala, S. R., Kumara, S. R. T., Rao, C. R., & Albert, R. (2013). Clustering social networks using ant colony optimization. *International Journal on Operational Research, 13*(1), 47–65.

Meisel, M., Pappas, V., & Zhanga, L. (2010). Taxonomy of biologically inspired research in computer networking. *Computer Networks: The International Journal of Computer and Telecommunications Networking, 54*(6), 901–916.

Mirollo, R. E., & Strogatz, S. H. (1990). Synchronization of pulse-coupled biological oscillators. *SIAM Journal on Applied Mathematics, 50*(6), 1645–1662.

Nakano, T. (2011). Biologically inspired network systems: A review and future prospects. *IEEE Transactions on Systems, Man, and Cybernetics, Part C: Applications and Reviews, 41*(5), 630–643.

Pawson, T. (1995). Protein modules and signalling networks. *Nature, 373*(6515), 573–580.

Srinivasan, M. V. (2011). Honeybees as a model for the study of visually guided flight, navigation, and biologically inspired robotics. *Physiological Reviews, 91*(2), 413–460.

Taniguchi, Y., Wakamiya, N., & Murata, M. (2006). A distributed and self-organizing data gathering scheme in wireless sensor networks. *International Journal of Wireless and Mobile Computing, 2*(1), 113–120.

Tatar, N., & Holban, S. (2012). A bio inspired alternative to Huffman Coding. In *Proceedings on Development and Application Systems* (Vol. 37, pp. 179–182).

Tschudin, C. F. (2003). Fraglets–A metabolistic execution model for communication protocols. In *Proceedings on Annual Symposium on Autonomous Intelligent Networks and Systems* (Vol. 6, No. 2, pp. 1–3).

Wang, M., & Suda, T. (2001). The bio-networking architecture: A biologically inspired approach to the design of scalable, adaptive, and survivable/available network applications. In *Proceedings on Symposium on Applications and the Internet* (pp. 43–53).

Wang, H., Zheng, R., Li, X., & Liu, D. (2006). A bio-inspired multidimensional network Security model. *In Computer and Computational Sciences, IMSCCS'06. First International Multi-Symposiums on* (Vol. 2, No.1, pp. 3-7).

Yu, F. R., Huang, M., & Tang, H. (2010). Biologically inspired consensus-based spectrum sensing in mobile ad hoc networks with cognitive radios. *IEEE Network, 24*(3), 26–30.

Chapter 2
Computer Networks

Abstract A computer network, in general, comprises of numerous computers that are linked together to communicate with each other. The goal of a computer network is to enable two or more computers to share and exchange data with one another for various purposes. Users can access remote resources by either logging into the appropriate remote computer or transfer data from the remote computer to their own computers. To understand what a network is all about, this chapter provides details on topologies, design, and usage of a network. Furthermore, since present network demands future technologies to be self-adaptive and self-healed, the chapter provides details on issues and challenges faced by it. Additionally, the chapter provides ground details on the future of networking technologies.

2.1 Introduction

As individual microchips in high-performance computing systems reach evermore prominent speed, execution starts to depend less on the rate of processors and more on the framework that supplies them with information. This framework is the network system, a regularly overlooked yet essential piece of any complex computer (Osborn 2015). Network system administration is to a great degree a wide subject in data science which calculates itself regarding profundity and significance. Taken just, on the other hand, a network system is close to a framework by which one can handle components and send data to another. Accordingly, the discriminating parameter of a network system is to measure the data stream. One essential metric is the transmission capacity or *bandwidth*, or the greatest rate at which a system can move data over a line that partitions the hubs/nodes into two equivalent gatherings. Generally, as essential for firmly coupled multiprocessing, the time needed to exchange a message between hubs is called the *latency*. An extensive and consistent examination exertion devotes itself to enhance these two numbers, bringing about a tremendous scope of way to deal with network system outline.

A network, in general, is formed by a collection of people, devices, and agents where the agents communicate with each other to share and gain resources.

A computer network is a collection of two or more devices connected for the purpose of sharing resources. Devices can include computers, printers, fax machines, and Internet communication hardware. In addition to hardware devices, software is also used to provide additional capabilities such as security (privacy and protection of network traffic) and enhanced services such as Internet browsing, print queue management, etc. These devices can be connected through wires (cables) or wireless technologies (radio or infrared).

2.2 Network Topologies, Types, and Design Strategies

While designing a network structure, network topologies, type, and design strategies play a very important role. Customarily, numerous individuals have thought that it was valuable to separate all the topologies into two vast gatherings, saying that a system is either static or dynamic. This is to some degree deluding on the grounds that not many genuine topologies submit to such simple characterization, rather falling some place on an expansive range in the middle of static and dynamic. A static system topology is supposed on the grounds, that it tries to give perpetual information ways that fit the correspondence needs of the application. In a consummately static topology, the system itself is a sensibly inactive gadget that essentially gives a gathering of channels with known and true endpoints. The perfect static system is a totally joined chart, where a committed channel exists in the middle of every single pair of hubs. For this situation, the system does no routing, and the sending node places the message on the right channel or the destination node. Involving the inverse end of the range are dynamic systems, which make altogether different suppositions about the system's part in taking care of messages. Dynamic systems endeavor to utilize adaption and intervention to share a littler number of physical connections among more nodes. A basic transport topology is a bus topology, which has precisely one physical channel that is shared by every single appended node. Given that both static and dynamic procedures have distinct tradeoffs, most real frameworks pick some all-around adjusted center ground. The following section presents details on various network topological designs and types.

2.2.1 Network Topologies

Topology refers to the shape of the network or in other terms it is the network layout. The way the computers in a network are physically linked to each other and how they communicate with each other is determined by the network topology as shown in Fig. 2.1. Topologies are either physical or logical such as:

(a) *Mesh Topology*: The simplest network connecting two computers A and B is an electrical link directly from one to the other. In a mesh topology, every

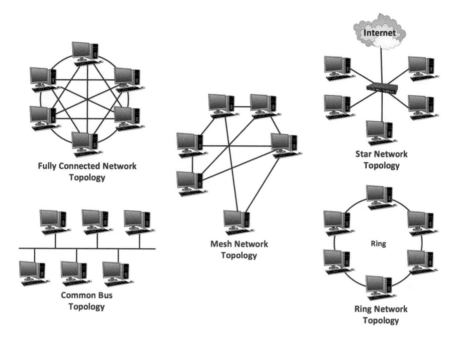

Fig. 2.1 Different network topologies

computer has a connection to every other computer in the network. Each computer has a network interface card (NIC) with a transmitter and a receiver. A packet is transmitted by one computer as a sequence of bits and received by the other in the same order. Depending on the times it takes to transmit one bit, the link has a capacity of bandwidth or bits per second. Mesh topology is preferred where dedicated connection is required and time is more important than infrastructure, cost of laying, and maintenance of physical or wireless media. This type of interconnection enables every computer to have $(N - 1)$ NICs and $N(N - 1)/2$ total number of links, where N is the number of computers.

(b) *Star Topology*: All the devices are connected to a central hub. Nodes communicate across the network by passing through the hub. To reduce the number of links, every node is connected to one central node in the star topology. A packet between any two nodes may need two hops via the central node. If the central node is the source or the destination, only one hop is required. The central node is a single point whose failure renders the entire network inoperative. In cases, where most communication is between one server and its clients, the star topology is especially useful.

(c) *Ring Topology*: The single point of failure can be avoided using the ring or bus topology. In the ring, every device is attached to a circular cable, so that each device is connected directly to two other devices, one on either side of it. As a packet circulates around the ring, every node can receive it. To avoid packet circulating indefinitely, the transmitting node removes it after one round. In order to handle these functions, the attachment is through an electronic circuit called a transceiver.

(d) *Bus Topology*: In the bus topology, all nodes are connected to a single cable, called the bus or backbone, with no active devices. All the nodes in the system are directly connected to that link (the bus), which may be organized as a straight line. The sites can communicate with each other directly through this link. Each node is connected to it by a single tap. This can be very reliable and inexpensive. The failure of one side does not affect communication among the rest of the sites. However, if the link fails, the network is partitioned completely. The bus is the most popular topology for local area networks (LAN) due to its simplicity and reliability. Initially, the most widely used LAN Ethernet uses the bus topology. Like the ring, the bus also has the broadcast property, i.e., as a packet propagates down the bus to its ends, it can be received by every node. In the mesh and star topologies sending the same packet to every node requires that N − 1 copies of the packet be separately transmitted.

(e) *Tree Topology*: Tree topology can be derived from the star topology. Tree has a hierarchy of various hubs, like we have branches in a tree. In this case, every node is connected to some hub or switch.

2.2.2 Network Types

The network computers may be linked through cables, telephone lines, radio waves, satellites, or infrared light beams. There are three basic types of computer networks:

- Local area networks (LAN)
- Metropolitan area network (MAN)
- Wide area network (WAN)

(a) Local Area Network

A local area network is normally a privately owned network within a single office, building, or campus covering a distance over a few kilometers. In a typical LAN configuration, one computer is designated as the file server. It stores all the software that controls the network as well as the software that can be shared by the computers attached to the network. Computer connected to the file server are called as workstations. In most of the LANs, cables are used to connect the NIC in each computer. Most LANs connect workstations and personal computers. Each node (individual computer) in a LAN has its own CPU with which it executes programs, but is also able to access data and devices anywhere on the LAN. The users can share expensive devices such as laser printers, communicate with each other by sending emails, or engage in chat sessions. The following characteristics differentiate one LAN from another:

- *Topology*: The geometric arrangement of devices on the network. For example, devices can be arranged in a ring or in a straight line.
- *Protocols*: The rules and encoding specifications for sending data. The protocols also determine whether the network uses peer-to-peer (P2P) or client–server architecture.
- *Media*: Devices can be connected by twisted-pair wire, coaxial cables, or fiber optic cables. Some networks communicate without connecting media altogether, instead, doing so through radio waves.

LANs are capable of transmitting data at very fast rates, much faster than data that can be transmitted over a telephone line, limitation being the number of computers attached to a single LAN. A LAN can be configured either as a client–server LAN or a P2P LAN as shown in Fig. 2.2.

Peer-to-Peer Model: P2P networks are the simplest and least expensive networks to set up. P2P networks are simple in the sense that the computers are connected directly to each other and share the same level of access on the network. Computer A will connect directly to computer B and will share all files with the appropriate security or sharing rights. If many computers are connected, a hub may be used to connect all these devices.

Client-Server Model: The most common LAN types used by companies today are the "client-server model," since they consist of the server (storing the files and running applications) and the client machines (computers used by the workers). Using a client–server setup can be helpful in many ways. It can free up disk space by providing a central location for all the files to be stored ensuring that the most recent file is available to all. A server can act as a mail server (collecting and sending the mails) or a print server (performing print jobs), thus freeing computing power on the client machine to continue working.

(b) Metropolitan Area Network

A metropolitan area network (MAN) covers larger geographical areas such as cities or districts. A system of LANs connected through telephone lines and radio waves is called as MAN. The connectivity lies among cities or districts where cities cannot lay a private network all around in the city.

Fig. 2.2 Client–server model versus peer-to-peer Model

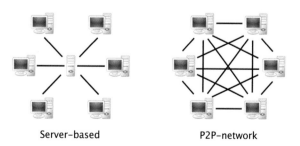

Server-based   P2P-network

(c) Wide Area Network

Wide area networks (WANs) are huge compared to a LAN or a MAN and span
across cities, state, country, continent, or even the whole world. WANs connect
larger geographical areas such as India, the United States, or the world. The satellite
uplinks may be used to connect this type of network. A WAN provides
long-distance transmission of data, voice, image, and video information. Using a
WAN, people in India can communicate with places like Tokyo in a matter of
minutes, without paying enormous phone bills. WAN technologies use multiplexers
to connect local and metropolitan networks like the Internet.

(d) Internetworks

When two or more networks are connected, they become an Internetwork, or
Internet. Individual networks are joined into Internetworks by the use of
Internetworking devices. These devices include routers and gateways.

2.2.3 Design Strategies for Communications

A network allows one computer to send electrical signals to another computer.
These signals have to be interpreted as a stream of bits. The stream of bits has
different meaning depending on the application. If one computer sends a binary file
and the other expects to receive an e-mail message, clearly communication will not
take place. In order to communicate, both parties must agree on a set of conversions
such as signals that constitute a 1 or a 0, such a set of conventions is a *protocol*.
    A network may include computers manufactured by different vendors with
software from various sources. For these to be able to communicate, the protocols
must be agreed upon by all the manufacturers, i.e., *standards* are required. For
instance, RS-232C is a standard protocol for transmission of a stream of bytes that
is widely used for sending data between computers and peripherals such as printers
and modems. TCP is a standard protocol used for reliable transmission of arbitrary
data between computers in the Internet.
    While designing a communication network, the systems on the networks agree
on a protocol or a set of protocol for determining host names, locating hosts on the
network, establishing connections, and so on. The design strategy should be sim-
plified by partitioning the problem into several layers. Each layer on one system
communicates with the equivalent layer on the other system. Each layer has its own
protocols or logical segmentation. The protocols may be implemented in hardware
or software. The logical communication between two computers can be implemented in three layers. The lowermost layer defines the electrical characteristics of
the link, the representation of buts, and the mechanical details of connectors and
cable. The middle layer handles the sending and reception of packets. At the highest
layer is the protocol for actually transferring the complete file. Layering of protocols
serves another purpose: it is possible for different vendors to implement different

layers and for these to interoperate provided they all conform to a standard layering. One such standard for layering is the International Standards Organization (ISO) reference model for Open Systems Interconnection (OSI).

OSI Reference Model

The OSI reference model provides a means of understanding the fundamentals of networking. It describes how data flows across a network. It helps in making better decisions about equipment purchases and configurations. It has the following seven layers with their basic tasks described below (Fig. 2.3):

Application layer: The application layer is responsible for a set of functions commonly required by various applications. It uses a set of protocols for carrying out these functions. Examples of some application layer protocols include HTTP, FTP, TELNET. Virtual terminal emulation and similar functionality are included in the list of responsibilities of the application layer.

Presentation Layer: The presentation layer is responsible for another set of functions commonly required by various applications sitting in the application layer. It is primarily concerned with the syntax and semantics of the information required to be transmitted over the network. Examples of some of the activities of this layer include data or information encoding in a manner a priori agreed between the sending and receiving parties. For instance, an ASCII system talking to an EBCDIC system may use services of this layer.

Session Layer: The session layer is responsible for establishment of one or more sessions between two or more users or applications working on different member systems of a network. Examples of some of the activities of this layer include selective flow of traffic in either directions, session control by token circulation, and synchronization of status information, etc.

Fig. 2.3 The OSI reference model

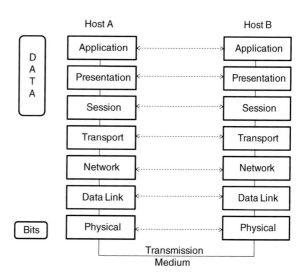

Transport Layer: The transport layer is responsible for receiving the data from the upper layer, i.e., the session layer and dividing it into manageable chunks for the purpose of further processing and onward transmission to network layer after prefixing its own header to the processed data. At the other end, this operation is reversed when this layer receives data from the network layer and after due processing passes it on to its upper layer. Other activities of this layer include creation of network connections as per the transport connection requests by the upper layer.

Network Layer: The network layer is responsible for receiving the data from the transport layer, process it for finding out the required resources, and divide the data into fragmented units. Thereafter, decide the route to be taken by the respective data units and pass the data to the lower (data link) layer after prefixing its own header to it. At the other end, this operation is reversed. Routing decision can be based on a fixed or static routing policy or a dynamic (situation dependent) routing policy. Other functions of the network layer include congestion control, address resolution, protocol translation, and resource usage accounting.

Data Link Layer: The data link layer is responsible for receiving the data from the network layer, process it, insert the processed data into data frame, add control information to it by prefixing a header, and suffixing a trailer to the processed data to pass it on to the physical layer for actual transmission in signal form. Examples of some of the other activities of this layer include data link layer protocol translation in required cases, ensuring acceptably error-free transmission, flow control, traffic direction regulation, and media access control in case of shared media systems.

Physical Layer: The physical layer is responsible for receiving the data from the data link layer, converting it into equivalent signal (representing the data in bits), and transmitting these signals in the desired manner over a shared or dedicated transmission link. Apart from the electrical characteristics, this layer is also concerned with the mechanical issues like connector dimensions, interpin distance, mechanical strength needed, etc. Issues like physical connection establishment, direction of transmission, frequency usage, and other procedural matters are under its purview.

2.3   Wireless Networking

Although the origins of radio frequency-based wireless networking can be traced back to the University of Hawaii's ALOHANET research project in the 1970s, the key events that led to wireless networking becoming one of the fastest growing technologies of the early twenty first century have been the ratification of the IEEE 802.11 standard in 1997, and the subsequent development of interoperability certification by the Wi-Fi Alliance (formerly WECA).

From the early 1970s to the early 1990s, the growing demand for wireless connectivity could not be met by a narrow range of expensive hardware, based on proprietary technologies, which offered no interoperability of equipment from different manufacturers, no security mechanisms, and poor performance compared to the standard 10 Mbps wired Ethernet.

The 802.11 standard stands as a major milestone in the development of wireless networking and the starting point for a strong and recognizable brand Wi-Fi. This provides a focus for the work of equipment developers and service providers and is as much a contributor to the growth of wireless networking as the power of the underlying technologies.

With the various Wi-Fi variants that have emerged from the original, 802.11 standard have grabbed most of the headlines in the last decade; other wireless networking technologies have followed a similar timeline, with the first IrDA specification being published in 1994, the same year when Ericsson started research on connectivity between mobile phones and accessories that led to the adoption of bluetooth by the IEEE 802.15.1 Working Group in 1999. Various 802 working groups are shown in Table 2.1 (Wikipedia 2015).

Table 2.1   IEEE 802 standards

Number	Topic
802.1	Overview and architecture of LANs
802.2	Logical link control
802.3	Ethernet
802.4	Token bus (was briefly used in manufacturing plants)
802.5	Token ring (IBM's entry into the LAN world)
802.6	Dual queue dual bus (early metropolitan area network)
802.7	Technical advisory group on broadband technologies
802.8	Technical advisory group on fiber-optic technologies
802.9	Isochronous LANs for real time applications
802.10	Virtual LANs and security
802.11	Wireless LANs
802.12	Demand priority (Hewlett Packard's AnyLAN)
802.14	Cable modems
802.15	Personal area network (Bluetooth)
802.16	Broadband wireless access (WiMAX)
802.17	Resilient packet ring
802.18	Technical advisory group on radio regulation
802.19	Technical advisory group on coexistence
802.20	Mobile broadband wireless access
802.21	Media independent handoff
802.22	Wireless regional area network
802.23	Emergency services working group

2.4   Usage of Networking

In general, computer communication and networking enables sharing of peripherals and resources which may be data, information, and supports communication among people. Some common usage of networks lies in remote login, file access, electronic mail, information access, remote printing.

(a) *Remote Login and File Access*: Network systems allow a user to issue commands for accessing remotely located computer. Furthermore, a user from any of the several client computers can read or modify files on single server computer. Thus, a group of people working on a common task can easily share data, documents, and programs.

(b) *Remote Printing*: The network enables use of a remote printer from any other computer without physically moving either the printer or the files.

(c) *Electronic Mail*: It enables a fast and convenient alternative to traditional post and telephone for inter person communication.

(d) *Information Access*: It ensures searching of databases which are available on remote machines. The development of World Wide Web (WWW) is one of the rapidly growing technologies of present network systems.

(e) *Storage Capacity*: Since there is more than one computer on a network system which can, without much of a stretch, offer documents, the issue of capacity limit gets set out all the things considered. A stand-alone computer may miss the mark concerning stockpiling memory, however, when numerous computer are on a network system, the memory of diverse computer can be utilized as a part of such case. One can likewise outline a stockpiling server on the system keeping in mind the end goal to have an immense stockpiling limit (Buzzle 2015).

(f) *Resource Sharing*: Resource sharing is another vital advantage of a computer network. For instance, if there are 12 representatives in an association, every individual has their own particular computer; they will oblige 12 modems and 12 printers on the off chance that they need to utilize the assets in the meantime. A computer network gives a less-expensive option by the procurement of asset sharing. Each computer can be interconnected utilizing a system and only one modem and printer can effectively give the administrations to each of the 12 clients (Kozierok 2015).

(g) *Economical Setup*: Shared assets mean decrease in equipment costs. Shared documents mean decrease in memory prerequisite, which in a roundabout way implies lessening in record stockpiling costs. A specific programming can be introduced just once on the server and made accessible over every single associated computer without a moment's delay. This spares the cost of purchasing and introduces the same programming the same number of times for the same number of clients.

(h) *Performance Enhancement and Balancing*: Under a few circumstances, a system can be utilized to upgrade the general execution of a few applications by conveying the calculation errands to different computer on the system.

Besides all the above-mentioned advantages, networking can be utilized in resolving various issues which the single stand-alone system cannot do.

2.5 Challenges and Issues of Networking

The development of network systems has encountered a few noteworthy steps, and the research center of every stride has been continuously changing and advancing, from ARPANET to OSI/RM, then high speed networking (HSN) and high-performance computing (HPN) (Gu and Luo 2006). Amid the development, network systems have gained incredible ground and increased extraordinary achievement. On the other hand, with the appearance and escalation of tussle, alongside the three troublesome issues (service customizing, resource control, and user management) of the current system, it is found that conventional Internet and its building design no more meet the prerequisites of cutting edge system. In this way, it is the next generation network which the present Internet must develop to. With the mentality of accomplishing significant direction for exploration on next generation system, this section breaks down a few quandaries confronting the current system situation.

2.5.1 Quality of Service

Quality of service (QoS) is a dynamic, broad topic which has significant origination since appearance. It is an imperative metric and a requirement parameter set comprising of data transfer capacity (bandwidth), delay, jitter, packet loss ration, and nature of voice/video. While the customary Internet can just give best effort transmission, the necessities of brilliant multimedia (voice and video) transmission cannot be fulfilled. Consequently, it prompts the significance of the powerful QoS ensured. IETF has advanced numerous administration models and components to meet distinctive QoS necessities, for example, IntServ/RSVP, DiffServ, traffic engineering, and QoS-based routing.

IntServ gives three sorts of administration models: *guaranteed service, controllable-load service*, and *committed rate service*. Notwithstanding, IntServ cannot be deployed on a vast scale on the grounds that each system hub needs to store numerous streams state (Xiao and Ni 1999). In this manner, DiffServ is advanced to defeat the confinement of IntServ, which is to give basic, adaptable, and separated administrations in Internet (Nichols et al. 1998; Carlson et al. 1998). However, DiffServ does not simplify QoS-guaranteed, but rather in the interim gives relative QoS-guaranteed to stream flow. The two administrations, IntServ and DiffServ, supplement and support one another in diverse applications. IntServ can be sent at the system edge or get to network to satisfy adaptable admission control and resource reservation. DiffServ works as a spine system of human body to satisfy productive information transmission.

Traffic engineering ensures QoS-guaranteed administrations by solving the unbalanced activity due to traffic circulation issue brought by routing protocol, and in addition the blockage issue created by improper resource utilization, hence giving QoS-guaranteed administrations to benefits. Likewise, QoS-based routing can fulfill distinctive QoS necessities (Mazumdar et al. 1991). In QoS-based routing system, path selection depends on QoS necessity of the accessible resource and data flow, which offers route of end-to-end limitation for more flows. QoS-based routing takes care of the issue that, path selection is just taking into account, single metric without considering the accessibility of resources (Crawley et al. 1998). Also, route change can help to adjust on the single system connection and enhance the productivity of system resource usage (Chen and Nahrsted 1998).

2.5.2 Connectivity, Manageability, and Scalability

Performance degradation alludes to issues including loss of speed and information uprightness because of poor transmissions and connectivity (IT Direct 2015). While each network system is inclined to execution issues, extensive systems are particularly powerless because of the extra separation, endpoints, and extra devices at midpoints.

Solutions for performance degradation are not appallingly troublesome. The primary step is to buy the best quality computer hardware equipment one can manage. Every single other arrangement expands upon a strong establishment of good system equipment. Of all the things considered, network performance is just tantamount to the parts of which it is created.

Albeit quality matters, for this situation, scalability can likewise be an issue. Systems without enough routers, switches, and bridges are practically identical to pumping water from a city well with a straw. Starting with satisfactory, quality equipment is an important aspect, however, that still is insufficient. Equipment is pointless without fitting setup.

It is crucial to guarantee all computers and system "pipes" are appropriately associated (with quality cabling) and arranged properly. This incorporates confirming system settings in server and desktop system design applications furthermore checking settings in the firmware of systems administration parts (switches, routers, firewalls, etc.). Each device joined on the system ought to be at first and routinely checked for issues, as computers tainted with viruses, spyware, and malware can squander data transmission and surprisingly more dreadful, contaminate different frameworks.

2.5.3 Network Security

Network system security issues include keeping up system integrity, keeping unapproved clients from invading the framework (survey/taking delicate information, passwords, and so forth.), and ensuring the system denial of service attack.

These issues are significantly amplified as systems increment in size. Bigger network systems are more helpless to assault in light, of the fact that they offer more powerless focuses at which interlopers can obtain entrance. More clients, more passwords, and more equipment mean more places, an intruder can attempt to get in. Barrier against these issues incorporate utilizing firewalls and proxies, introducing solid antivirus programming, installing strong antivirus software, making utilization of system investigation programming, physically securing computer organizing resources, and summoning methods that compartmentalize an extensive system with inside limits. These three issues, as comprehensively including as they may be, can be overpowering for little to average-sized business to handle all alone.

2.5.4 Network Congestion

Numerous system interferences connected with network demands are identified with signaling overload and data transmission overburden/bandwidth utilization. It is imperative to comprehend and have the capacity to scale and improve the signaling in the network system guaranteeing that the unlimited number of devices and applications are not bringing on pointless clog or in other terms, network congestion. Spectrum is the most profitable resource in the network domain. Spectrum increments are basic to network scope and to have the adaptability to meet the regularly extending number of users and data transmission requests. An expanded and well-utilized spectrum will convey better client experience. Measured performance execution of the system sign is imperative to the client experience. Network planning and tuning can convey up to three times the change in bandwidth. Calibrating the system through expert administrations is generally as critical as the equipment. Moreover, reconfiguring the systems rapidly (utilizing virtualization and a software-defined networking methodology) to rapidly test new administrations can rapidly scale it to millions. There is no more the need to contribute months of operational arranging to trial and offer service. The OpenDaylight project has given an open-source stage where individuals team up to build common software-defined networking infrastructure. Ericsson's commitment to the joint effort is centered around extending this insight from the data center into the network system—giving a more incorporated and less-difficult improvement environment.

2.6   Future of Networking

The era of modern business systems started more than a quarter century and was stamped by the public packet mode network as an alternative option for leased line-based wide area networks (WANs). Frame Relay formed the first epoch of this new era. It rose to prominence in the mid-1990s when organizations were saddled with numerous, merchant particular systems supporting centralized computer and customer/server situations. Frame Relay was intended to be convention straightforward and utilized supplant-isolated leased line WANs with a single multiprotocol packet network.

Before the end of the 1990s, Microsoft-controlled PCs and LANs were pervasive crosswise over organizations pushing out other exclusive network systems. At the same time, the Internet was prospering and TCP/IP turned into the overwhelming route to join branch, campus, and datacenter LANs together. The second age of cutting edge organizing, i.e., LAN internetworking was borne.

Today, IP-based devices are quickly multiplying inside and outside the four dividers of organizations making systems progressively hard to scale, manage, secure, and adjust.

Likewise, with most engineering issues, all aspects of system configuration present tradeoffs. It is troublesome, hence, to focus the essential heading of current network system research, on the grounds that it is assaulting numerous fronts at the same time. Regardless, there is adequate opportunity to get better on advanced system frameworks. Indeed, even the quickest systems are significantly slower than as far as possible forced by the rate of light, and enhancements in this idleness would have an awesome impact on multiprocessing execution. There is likewise the likelihood that new strategies for parallelizing applications will lead the network organization in already unexplored headings. Regardless, networking technology is in no way, shape or form a depleted science, and much work stays to be finished.

When we glance back at every age of cutting edge organizing, we see a watch's change at every move. Network technology has shown phenomenal growth in the today's global developing scenario. The network system should be adaptive and healed in such a way that if there are any catastrophic errors at one end, it should not affect the overall network performance. The network should show minimal errors with maximum output.

2.7   Summary

The union of computing and network systems administration is more apparent than in the amazing development of the WWW. In another sense, however, network systems administration is being pulled in two inverse directions. From one perspective, the Web's prominence and development have been powered to a great extent by desktop applications expanding transfer speed concentrated pictures and videos. From the other perspective, thin-client computers are turning out to be all the more generally utilized as edge-of-network system devices, frequently associated by wireless technology. There is likewise an expanding befuddling between fiber-optic transmission data transfer capacities and computer speed, pushing the processing further far from the system center. Based on it, this chapter provides details on network structure, type, and topology and the issues pertaining to the present network scenario. Furthermore, the chapter provides insights on the issues and challenges of networking.

References

Buzzle. (2015). *Advantages and disadvantages of computer networks.* http://www.buzzle.com/articles/advantages-and-disadvantages-of-computer-networks.html. Accessed September 29, 2015.

Carlson, M., Davies, E., Nortel, U. K., Wang, Z., & Weiss, W. (1998). *An architecture for differentiated services.*

Chen, S., & Nahrsted, K. (1998). An overview of quality of service routing for next-generation high-speed networks: problems and solutions. *Network IEEE, 12*(6), 64–79.

Crawley, E., Sandick, H., Nair, R., & Rajagopalan, B. (1998). *A framework for QoS-based routing in the internet.*

Gu, G. Q., & Luo, J. Z. (2006). Some issues on computer networks: Architecture and key technologies. *Journal of Computer Science and technology, 21*(5), 708–722.

IT Direct. (2015). http://www.gettingyouconnected.com/the-top-3-issues-affecting-todays-large-computer-networks/. Accessed September 28, 2015.

Kozierok, C. M. (2015). http://www.tcpipguide.com/free/t_TheAdvantagesBenefitsofNetworking.htm. Accessed September 29, 2015.

Mazumdar, R., Mason, L. G., & Douligeris, C. (1991). Fairness in network optimal flow control: Optimality of product forms. *IEEE Transactions on Communications, 39*(5), 775–782.

Nichols, K., Black, D. L., Blake, S., & Baker, F. (1998). *Definition of the differentiated services field (DS field) in the IPv4 and IPv6 headers.*

Osborn, C. (2015). *A networking overview.* http://www.ai.mit.edu/projects/aries/course/notes/networkpaper.pdf. Accessed September 29, 2015.

Wikipedia. (2015). https://en.wikipedia.org/wiki/IEEE_802. Accessed September 29, 2015.

Xiao, X., & Ni, L. M. (1999). Internet QoS: A big picture. *Network IEEE, 13*(2), 8–18.

Chapter 3
Inceptive Findings

Abstract Initially, mapping of biological systems with network systems was made using different protocols and rules that were derived structurally. The structural analogy is required for building a framework between the two systems. Looking for specific behavior that optimally maps to networks provides better understanding for modeling and systematic development of novel methods in network systems. Instigation of novel methodologies requires identification of the analogies and understanding of the realistic biological behavior. This chapter deals with the structural and operational correlation between biological systems and network systems. Initially, it discusses how the network domain can be compared to biological systems. Later it discusses the models developed in this regard.

3.1 Introduction

In a sense, the whole history of computer science is the history of a series of continuous attempts to discover, study, and implement computing ideas, models, and design paradigms (Paun 2005). The evolution of networking technology has brought many potential changes in our daily lives. Existing systems have shown tremendous development in terms of effective management, functionality, and articulating the design architectures with limited technical flaws and faults. However, next-generation networking architectures have increased the complexity and dynamicity with a high level of failures, which cannot be solved and handled by conventional methods. Challenges such as scalability, connectivity, complexity, and manageability require novel technology for designing, handling, and engineering network systems. The evolution of natural computing has yielded artifacts and has provided enough efficiency and elegance in controlling various issues and challenges since decades. Nature and network technology goes hand in hand in demands of resource management, self-adaptability, and self-healing characteristics. Therefore, for the development and evolution of networking technology, many researchers are engaged in investigating innovative paradigms to address various challenges in computer network technology.

The general approach to develop bio-inspired networking technology is to discuss the identification of biological structures and techniques relevant to network systems followed by, modeling the system characteristics and providing engineering solutions for optimization. Mapping the biological systems to the network systems is the first and foremost task in developing bio-inspired models. Mapping the two structures is investigated by identifying the analogies between the two systems. In developing any bio-inspired model, it is crucial that the structures and methodologies between the two systems should be similar.

3.2 Structural Composition

A network in general is composed of a vast network domain that has sub-networks followed by network nodes in each sub-network, as shown in Fig. 3.1.

Similarly, a biological organism carries out basic life processors that are required for healthy living through a set of organ systems. The biological system is subdivided into organs that are made up of tissues. A tissue in turn has cells that are the basic unit of living bodies. Structurally, network domains can be compared to organs, sub-networks to tissues, and network nodes to cells.

Cells are the basic unit of biological organisms. They provide the specificity for information transfer in the living bodies. Cells communicate with each other in a manner similar to how nodes communicate with each other (Elsadig and Abdullah 2008). From a local point of view, information transfer works as a signal that reaches neighboring cells. Signaling in living bodies occurs in two ways, intercellular and intracellular. Intracellular processing occurs in the single cell entity to achieve cellular response via initiating complex cascades. The intracellular processing of cell initiates when a receptor is transferred on the cell surface. Later, the

Fig. 3.1 A network domain comprising subnetwork and network nodes

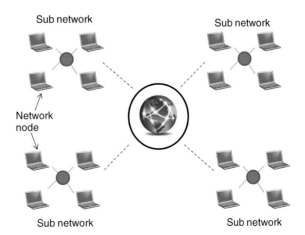

reorganization of cell structure happens to process the information for initiating a cellular answer. The cellular response can be of various types such as secretion of hormones, submission of molecule, etc.

Intercellular cell communication occurs among cells to reach the destination via signaling cascading. The signal induced by cell processing initiates a signaling cascade in each target cell resulting in a very specific answer, termed as response, as shown in Fig. 3.2. The signal is transferred into the bloodstream of biological organisms for reaching the specific destination. The adaptation of mechanisms known from intracellular and intercellular processing promises to enable more efficient information exchange. In networks, node to node transmission occurs in a similar manner where the information received is processed and likewise response is given to induce signal to the other node.

From a different perspective, the human body has its own functionality in protecting the body from foreign bodies that are injurious to it. The body has two major defensive and protecting layers, innate and adaptive layers that are discussed in detail in Chap. 4. Primarily, the human skin is toughened by collagen and a set of bacteria to protect us from pathogenic bacteria. On similar grounds, this protection can be compared to the function of firewall in networks, as shown in Fig. 3.3 (Elsadig and Abdullah 2008). As firewall controls the incoming and outgoing traffic from the network, a similar functionality is performed by the skin. Later, it has the interface and router which can be compared to the innate and adaptive immunity. Innate immunity is the first line of defense and adaptive is the second line of defense. Similarly, interface and router performs traffic management by analyzing and monitoring it.

Biological systems primarily function for information processing. Information processing in general requires computation, thought handling, memory storage, communication, and control system. The human brain can be compared to the heart of a computer program (Searle 1990). Analogy between the human brain and the computer system can be drawn in the following ways:

1. *Sensation*: The gathering of information in a computer system can be compared to sensations felt by our body with respect to various stimuli.
2. *Perception*: After data are gathered, useful information and structure is inferred from the received data. The human brain calls it perception, where the sensed information is understood and interpreted in various ways.

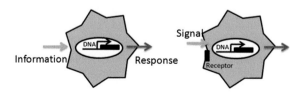

Fig. 3.2 Intercellular communication (*Source* Elsadig and Abdullah 2008)

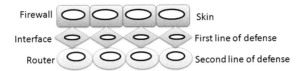

Fig. 3.3 Skin–firewall relationship (*Source* Elsadig and Abdullah 2008)

3. *Memory*: Next, the information is stored and recalled in the inbuilt storage capacity which is present in every computer system, i.e., random access memory (RAM), read only memory (ROM), and flash memory. The memory is required to process the data later if required. Both the human brain and the computer has inbuilt memory to store and process data.
4. *Decision*: Based on the information received, planning and management of data is made for guiding future actions. The decision is based on the information processed through the internal computation and control system.
5. *Behavior*: What we call behavior in social psychology with respect to the input received is called the output of the computer system. The behavior of humans is based on factors such as affection and cognition (McAllister 1995).
6. *Learning*: The knowledge acquired when the above sequence of tasks is carried out is an experience learned in the human brain. Similarly, the computer learns the consequences of the carried out actions.

In this manner, structural network systems can be compared to biological systems and analogy can be derived between the two systems.

3.3 Highly Optimized Tolerance (HOT) Model

Carlson and Doyle formulated the Highly Optimized Tolerance Model (HOT Model). HOT Model states that a complicated and strictly organized internal structure is necessary for any system to exhibit robust external behavior (Carlson and Doyle 1999).

The human body has many different organs and a physiological system, each serving a specific purpose. For example, the kidney maintains the acid–base balance to regulate blood pressure, a functionality that cannot be served by the lung. The lungs on the other hand, help us to breathe air to inhale oxygen. Red blood cells then carry the inhaled oxygen around the body to be used in the cells found in our organs and tissues. Lungs also help the body to get rid of CO_2 gas when we breathe out. In a similar manner, the Internet, a well-known computer network, contains a number of specialized devices with specific functionality. At its core are high speed routers which forward data in a highly optimized manner. At its edges are computers, each having its own functionality and cannot take over others' functionality. HOT model thus offers bases to bio-inspired systems.

3.4 Biological Evolution

The Internet evolution can be mimicked from the biological evolution, which has similarities with the following three facts (Dovrolis 2008):

- *Genetic Heritage*—In biological organisms, backward compatibility inherits the characteristics and features from previous generations. Comparably, in network systems, network offspring inherits most architecture, protocols, underlying technology, and applications of their predecessors.
- *Variation*—Genetic diversity occurs in such a way that the environment chooses the best mutation from the plethora of mutations. Correspondingly, there are specific changes in the existing network species at the architectural, protocol, technological, or application levels.
- *Natural Selection*—It is a process through which the environment chooses one among its mutations. Similarly, on certain occasions IPv4 is preferred over IPv6. The Internet has not always selected protocols or architectures that are well designed, like IPv6 over IPv4.

Another trivial example of linking the two fields is the hourglass model of IP. In biology, the hourglass model is similar to the IP hourglass model. Here, the animal morphology tends to be conserved during the embryonic phylotypic period which is a period of maximal similarity between the species within each animal phylum (Leigh 2007). As shown in Fig. 3.4, it leads to the proposition that embryogenesis diverges more extensively early and late than in the middle. Doyle and Csete (2011) emphasized it as the IP hourglass model of Internet architecture, i.e., a diverse and rapidly changing set of applications run on top of a smaller set of transport protocols, which in turn run on a single Internet protocol (Akhshabi and Dovrolis 2011).

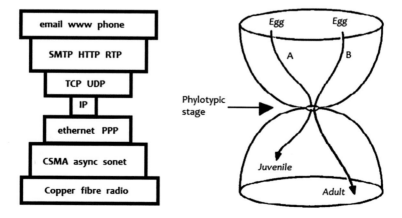

Fig. 3.4 Hourglass model of network and biology

3.5   Natural Computing

Researchers have developed highly robust, fault tolerant, adaptive, dynamic, and complex bio-inspired systems with extremely magnificent capabilities of delivering efficient systems that have revolutionized our understanding of the two systems. Neural computing, DNA computing, immune system computing, and evolutionary computing are some of the splendid examples of the present computation derived from the human body. Figure 3.5 shows the mapping between diverse components of the human body and the fields of bio-inspired computing.

The earliest of the nature-inspired computation models were the cellular automata given by Ulam and von Neuman in 1940 (Kari and Rozenberg 2008). Von Neumon expressed that cell automata is the structure for comprehension of the complex nature of systems. He additionally trusted that self-reproduction was an element crucial to both natural creatures and computers.

In parallel, ahead of correlations (Von Neumann and Kurzweil 2012) between computing machines and the human nervous system, McCulloch and Pitts (1943) proposed the very first model of artificial neurons. This exploration in the end offered ascends to the field of neural calculation and likewise had a significant impact on the establishment of automata hypothesis (Kleene 1951). Then again, it was anticipated that, by utilizing the standards of how the human cerebrum forms data, neural calculation would yield critical computational advances (i.e., how would we be able to fabricate a multipurpose computer?). The primary objective has been sought after, mostly inside of the neurosciences under the name of cerebrum hypothesis or computational neuroscience, while the mission for the second objective has turned out to be primarily a software engineering known as artificial neural or just neural systems.

While Turing and von Neumann longed for comprehension of the cerebrum, (Teuscher 2012) and perhaps outlining a smart computer that works alike the mind,

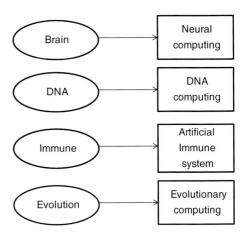

Fig. 3.5 Mapping human body with bio-inspired network models

evolutionary computation (Back et al. 1997) was raised as another calculation worldview that drew its motivation from a totally diverse piece of science: Darwinian evolution (Darwin 2009). As opposed to imitating components of a solitary natural life form, evolutionary computation draws its motivation from the motion of whole types of creatures. It includes a steady or variable-size populace of people, a wellness paradigm as per which the populace of people is being assessed, and hereditarily propelled administrators that deliver the cutting edge from the present one.

Artificial immune systems are computational frameworks contrived beginning in the late 1980s and mid-1990s (Farmer et al. 1986; De Castro and Timmis 2002), as computationally intriguing deliberations of the regular safe arrangement of natural living beings. Seen as a data preparing framework, the insusceptible framework performs numerous intricate calculations in an exceedingly parallel and disseminated design (DasGupta 1999). It uses learning, memory, affiliated recovery, and different instruments to take care of acknowledgment and arrangement issues, for example, qualification in the middle of self and non-self cells, and balance of non-self pathogens. Undoubtedly, the regular insusceptible framework has infrequently been known as the "second brain" (Bertalanffy 1951) on account of its capable data preparing capacities.

Molecular computing (referred additionally as biomolecular computing, bio-computing, biochemical computing, DNA processing) depends on the thought that information can be encoded as biomolecules, for example, DNA strands and sub-atomic science apparatuses can be utilized to change this information to perform, for instance, math or rationale operations. The conception of this field was the 1994 achievement test by Leonard Adleman, who fathomed the Hamiltonian path problem singularly by controlling DNA strands in test tubes (Adleman 1994).

3.6 Feedback Loops for Kidney-Blood Pressure

Kidney, a vital organ of the human body, has the main function of regulating blood pressure via maintaining the salt and water balance. Angiotensin-based regulation process for blood pressure was proposed by Dressler et al. (2005). They used the mechanism to model the control loop for an efficient regulatory process in an organism (Fyhrquist et al. 1995). There are a series of conversions and activations that happen when there is decrease in arterial blood pressure as shown in Fig. 3.6. Following are the series of activities when there is increase in blood pressure:

- *Renin–Angiotensin Mechanism*: As renin travels through the bloodstream, it binds to an inactive plasma protein, angiotensinogen, activating it into angio-tensin I.

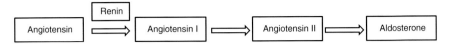

Fig. 3.6 Renin–angiotensin–aldosterone increases blood volume and pressure

- *Converting Angiotensin I to Angiotensin II*: As angiotensin I passes through the lung capillaries, an enzyme in the lungs converts angiotensin I into angiotensin II.
- *Angiotensin II Stimulates Aldosterone Release*: Angiotensin II continues through the bloodstream until it reaches the adrenal gland. ADH (made in hypothalamus and released by the posterior pituitary) is also simulated, which causes arteries to constrict.
- *Release of Aldosterone*: It stimulates the cells of the adrenal cortex to release the hormone aldosterone. As aldosterone is released, it causes nephron distal tubules to reabsorb more sodium and water, which increases the blood volume.

Here, renin is produced by the kidney, angiotensin is made by the liver, and aldosterone is produced by the adrenal glands located on top of the kidneys.

The authors in Dressler et al. (2005) argued that, in a sensor network, control of activities requires exchange of information between multiple nodes. It is similar to kidney arterial low blood pressure in terms of sequential activity, which occurs when there is a decrease in blood pressure. In the sensor network, control information must be transported to the appropriate destination. Also, the destination must respond to request by confirming the instructions. Density of the sensor network allows for alternate feedback loops via the environment:

- Directly via the physical phenomena which are to be controlled by the infrastructure.
- Indirect communication, which allows for more flexible organization of autonomous infrastructures and reduces control messages.

3.7 Summary

The structural composition of biological organisms can be noticeably compared with different computer network systems. It gives the strong basis as to why the two systems act in accordance with each other. The chapter explained the relevance of directly or indirectly linking the two systems by diverse means. The basic understanding lies in comparing the domains with biological ones and adapting in solving issues such as data transmission, information propagation, etc. The later chapters explain in detail various networking issues with biological solutions.

References

Adleman, L. M. (1994). Molecular computation of solutions to combinatorial problems. *Science, 266*(5187), 1021–1024.

Akhshabi, S., & Dovrolis, C. (2011). The evolution of layered protocol stacks leads to an hourglass-shaped architecture. In *Proceedings of ACM SIGCOMM.*

Back, T., Fogel, D. B., & Michalewicz, Z. (1997). *Handbook of evolutionary computation.* IOP Publishing Ltd.

Bertalanffy, L. U. D. W. I. G. (1951). Theoretical models in biology and psychology. *Journal of Personality, 20*(1), 24–38.

Carlson, J. M., & Doyle, J. (1999). Highly optimised tolerance: a mechanism for power laws in designed systems. *Physical Review E, 60,* 1412–1427.

Darwin, C. (2009). The origin of species by means of natural selection: Or, the preservation of favored races in the struggle for life. In W. F. Bynum, A. L. Burt (Ed.).

DasGupta, D. (1999). *An overview of artificial immune systems and their applications* (pp. 3–21). Berlin: Springer.

De Castro, L. N., & Timmis, J. (2002). Artificial immune systems: A new computational intelligence approach. Springer Science & Business Media.

Dovrolis, C. (2008). What would Darwin think about clean-slate architectures? *ACM SIGCOMM Computer Communication Review, 38*(1), 29–34.

Doyle, J. C., & Csete, M. (2011). Architecture, constraints, and behavior. *Proceedings of the National Academy of Sciences, 108*(Supplement 3) (pp. 15624–15630).

Dressler, F., Krger, B., Fuchs, G., & German, R. (2005). Self-organization in sensor networks using bio-inspired mechanisms. In *Proceedings of International Conference on Architecture of Computing Systems—System Aspects in Organic and Pervasive Computing (ARCS'05): Workshop Self-Organization and Emergence* (Vol.18, pp. 139–144).

Elsadig, M., & Abdullah, A. (2008). Biological inspired approach in parallel immunology system for network security. In *Proceedings of International Symposium on Information Technology* (Vol. 1, pp. 1–7).

Farmer, J. D., Packard, N. H., & Perelson, A. S. (1986). The immune system, adaptation, and machine learning. *Physica D: Nonlinear Phenomena, 22*(1), 187–204.

Fyhrquist, F., Metsärinne, K., & Tikkanen, I. (1995). Role of angiotensin II in blood pressure regulation and in the pathophysiology of cardiovascular disorders. *Journal of Human Hypertension, 9,* S19–S24.

Kari, L., & Rozenberg, G. (2008). The many facets of natural computing. *Communications of the ACM, 51*(10), 72–83.

Kleene, S. C. (1951). *Representation of events in nerve nets and finite automata* (No. RAND-RM-704). *Rand Project Air Force Santa Monica Ca.*

Leigh, S. R. (2007). Homoplasy and the evolution of ontogeny in papionin primates. *Journal of Human Evolution, 52*(5), 536–558.

McAllister, D. J. (1995). Affect- and cognition-based trust as foundations for interpersonal cooperation in organizations. *The Academy of Management Journal, 38*(1), 24–59.

McCulloch, W. S., & Pitts, W. (1943). A logical calculus of the ideas immanent in nervous activity. *The Bulletin of Mathematical Biophysics, 5*(4), 115–133.

Paun, G. (2005). Bio-inspired computing paradigms (natural computing). *Lecture Notes in Computer Science, 3566,* 155.

Searle, J. R. (1990). Is the brain's mind a computer program. *Scientific American, 262*(1), 26–31.

Teuscher, C. (2012). Turing's connectionism: An investigation of neural network architectures. Springer Science & Business Media.

Von Neumann, J., & Kurzweil, R. (2012). The computer and the brain. Yale University Press.

Chapter 4
Swarm Intelligence and Social Insects

Abstract Swarm intelligence is a behavior shown by a collection of social insects and animals which exhibit spatial arrangement and synchronized motion. These animals control and manage their position with the help of local interactions among companions of the same species. The work is performed in a systematic manner such that exchange of information occurs. The information which is gathered is shared with the other species. This ability benefits them in many aspects of social life, such as the need to protect themselves from predators and to perform well-organized locomotion and foraging (Nicole in Fish, networks, and synchronization 199:3518–3562, 2012).

4.1 Ant Colony Optimization

Ants collectively search for food which is their main task for survival. Ants, and while searching for food they can adapt themselves to the changes in the environment, optimizing the path between the nest and the food source. This is due to stigmergy which involves positive feedback, given by continuous deposit of chemical substance, known as pheromone (Beckers et al. 2000). Classic example of searching for food via the shortest path can be seen in Fig. 4.1. Ants converge to the shortest path by selecting the path which has a higher deposition of pheromone. Pheromone is the physical substance secreted by the ants. As seen from Fig. 4.1, initially, half of the ants take one path and the other half take another path to reach to the food source. On their way, they deposit pheromone. The ants which first reach the food source are the first ones to come back again and give this information to other ants. Hence, the amount of deposition in the shorter path would be higher (see Eq. (4.1)) (Dorigo and Gambardella 1997).

$$\Delta\tau_{ij}^{k} = \begin{cases} \frac{Q}{L_k} & \text{if ant } k \text{ used edge } (i,j) \text{ in its tour} \\ 0 & \text{otherwise} \end{cases} \tag{4.1}$$

where, $\Delta\tau_{ij}$ is pheromone deposited from node i to node j and Q is a constant. The amount of pheromone deposition is inversely proportional to the length of the path.

© Springer International Publishing Switzerland 2016
H. Rathore, *Mapping Biological Systems to Network Systems*,
DOI 10.1007/978-3-319-29782-8_4

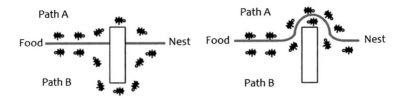

Fig. 4.1 Ants converging towards the shortest path (*Source* Perretto and Lopes 2005)

Longer the path lesser the pheromone deposition and shorter the path higher the pheromone deposition. As time passes, pheromone is updated using following equation (Dorigo and Gambardella 1997):

$$\tau_{ij} \leftarrow \tau_{ij} + \sum_{k=1}^{m} \Delta\tau_{ij}^{k} \qquad (4.2)$$

On a similar note, pheromone evaporates at the evaporation rate of ρ as depicted in the equation (Dorigo and Gambardella 1997):

$$\tau_{ij} \leftarrow (1 - \rho)\tau_{ij} \qquad (4.3)$$

Similar type of optimization can be applied in various routing algorithms (Leibnitz et al. 2006). There are various algorithms based on ant colony optimization namely AntNet, AntHocNet, GPS/Ant-like algorithm (GPSAL), ant-colony based routing algorithm (ARA), termite routing protocol, HOPNET.

Ant-based control (ABC) models destinations as food sources and routing tables as pheromone tables. It uses probabilistic approach to update its pheromone table. At a node x, each entry $(d, x \rightarrow y)$ in the pheromone table is the probability that an ant leaving node x will use link $x \rightarrow y$ to reach destination d. Symmetric cost on all links and reinforcement learning techniques are used in this technique.

In AntNet technique, ants take all paths with equal probability. There are two types of ants in this technique, viz., forward ants and backward ants. Forward ants collect the information by recording their path, and also the actual time when they arrive at each node. The backward ants update probabilistic table entries. This model supports asymmetric link costs. Two types of queues are maintained in this model, viz., low priority queue and high priority queue (Caro and Dorigo 1997). Ants reinforce the solution by the reinforcement parameter which is calculated by using trip times which they experience (see Algorithm 4.1). Unlike ABC, AntNet uses stochastic forwarding for data packets as well. The goal achieved by doing so is load balancing over multiple paths. However, this approach requires long delays to propagate routing information, since routing tables are only updated by backward ants.

AntHocNet technique is mainly used in MANETs. The technique bears a strong resemblance to the ad hoc on-demand distance vector (AODV) protocol, except for active probing and stochastic routing (Caro et al. 2005). It uses hello packets to detect link failures. The main strength of this technique lies in failure handling, since it takes advantage of the availability of multiple active paths.

GPS/Ant-like routing algorithm (GPSAL), used for routing in ad hoc networks, is based on global positioning system (GPS) and mobile software agents which are modeled on ants (Ciimara and Loureiro 2001). Ants are used to collect and disseminate information about the location of nodes in the network. This is a key aspect of the GPSAL algorithm that helps in accelerating route discovery. Another feature of this algorithm is the use of fixed hosts to route packets whenever possible.

The combination of these principles provides a better MANET routing algorithm. Ant-colony based routing algorithm (ARA) is distributed, loop-free and demand-based with a sleep period operation. Additionally, ARA has locality and multipath properties (Gunes et al. 2002).

Algorithm 4.1: AntNet Algorithm

Input:
s: source node
d: destination node
Output: Computes shortest path between source and destination node.

1 At regular intervals, from s, a mobile agent (forward ant), is launched.
2 Forward ants uses probabilistic routing tables together with queue status at every intermediate node to choose its output port from unvisited list of nodes.
3 Time elapsed and node identifier is pushed to ants stack.
4 if cycle detected then
5   delete cycle from ants' memory
6 if forward ant reaches to its destination then
7   transform itself to a backward ant
8   Visits the list of the nodes in its stack in a reverse order.
9   Update corresponding entries in the routing tables and array on its way back to source by using its values stored on its stack.

4.2   Bird Colony Optimization

When birds collectively flock together they exhibit a coordinated beautiful pattern. They follow a behavioral model in which each agent follows three rules:

- *Separation*: Each agent tries to move away from the neighbor if they are too close to equal separation among the neighboring agents.
- *Alignment*: They align either in a straight line or in a parallel formation.
- *Cohesion*: Each agent tries to position itself to the average position of neighbors.

The velocity and position of a particle or agent at time t can be represented by $v(t)$ and $x(t)$, respectively. This is made to select the best position and fitness as depicted in the following equations (Baguda et al. 2012):

$$v(t+1) = \omega v(t) + c1(pl + x(t)) + c2(pg + x(t)) \qquad (4.4)$$

$$x(t+1) = x(t) + v(t+1) \qquad (4.5)$$

where $v(t)$ keeps track of particle flight direction and prevents particle from sudden change in direction. $c1(pl + x(t))$ measures the performance of particle relative to past performance and $c2(pg + x(t))$ measures the performance relative to neighbors.

Since multimedia applications are sensitive to delay, higher priority must be given for the fact that they can be transported over unpredicted channel. In a video streaming, media is sent over a continuous stream of data which is placed on its arrival. Particle swarm intelligence is suitable because of its fast convergence and simplicity. The problem can be represented as

$$G = \{N, L\} \qquad (4.6)$$

where, N denotes the set of nodes and L denotes the set of communication link connecting the nodes together.

Algorithm 4.2: Video streaming based on Bird Colony

Input:
φ: optimal solution
Objective function: min $f(Q,D)$
Output:
Streamed video: $\varphi_{opt}= argmin\, f(Q,D)$
1 Initialize swarm with random positions and velocities of n particles
2 for all n particles do
3   Compute the values of Q_i, D_i at time t
4   Generate new velocity
5   Calculate new position based on the new velocity
6   **Comment**: Evaluate objective function
7   Check delay constraints if less than packet deadline
8   Compute the fitness of the particle based on:
        $f(Q,D)=af(q)+(1-a)f(t)$
9   Select local best
10   Select global best
11   if current fitness of particle is < previous and global then
12   replace the previous
13   Update particles velocities and positions
14 Increase the loop counter

S is the set of video source through which nodes stream the video content and $L(s)$ is the link used by the source s. For any source to stream the video content, the packets are queued first in buffer Q, before they are served. The queue size changes randomly and it can be represented as a set, where $Q = q_1, q_2, ..., q_n$. The delay anticipated after the source has placed a request for the video content is termed as D, where $D = d_1, d_2, ..., d_n$. The pseudo-code for the same is explained in Algorithm 4.2.

This approach has shown more promising results for achieving high convergence in time-sensitive video applications. It has proven to be very effective, simple, flexible, and high searching capability algorithm which can potentially be used in making critical decisions in error-prone transmission environment.

4.3 Bee Colony Optimization

Forage selection is defined as the act of looking or searching for food. Bee colony has three main components for forage selection.

- *Food source*: Value of food source depends on factors such as proximity to nest, richness or concentration of energy, and ease of extracting. There are two basic foragers which have a prominent role in taking nectar from flowers and in making a bee hive.
- *Employed foragers*: They are associated with a particular food source for which they are currently employed.
- *Unemployed foragers*: They are of two types, namely scouts and onlookers. Scouts are the ones which search for other food source in the surrounding environment, whereas onlookers wait in the nest and establish a food source through the information shared by employed foragers. Onlookers communicate with employed foragers with the help of a peculiar dance called as waggle dance in dancing area.

Algorithm 4.3: Forward pass of Huffman coding

Input: Set of bees
Output: Set of Huffman codes
1 for i=1 to bees do
2   if code not assigned (i) then
3   generate random number (i, 255)
4   while do
5   generate random mask ()
6   if mask is acceptable () then
7   break
8   assign code
9 call Backwardpass()

Huffman coding is used to encode symbols used for lossless data compression. Huffman coding using bee colony optimization algorithm was proposed in Tatar and Holban (2012). Their approach eliminated the drawbacks of using tree structure to generate Huffman codes in terms of lookup time. Huffman coding uses three concepts, i.e., the prefix property, minimality of the codes, and the frequency of the symbol, to generate codes. The algorithm inspired from bees is divided into two passes: forward pass and backward pass (see Algorithms 4.3 and 4.4, respectively).

The performance, in terms of lookup time, of this encoding technique is significantly better than the traditional structural approach.

Algorithm 4.4: Backward pass of huffman coding

Input: Set of bees
Output: Set of huffman codes
1 for i=1 to bees do
2   for j=1 to bees do
3   if code(i)<code(j) then
4   if prefix (code(i),code(j)==TRUE then
5   code(j)=0;
6 for i=1 to bees do
7   if code(i)==0 then
8   call Forwardpass()

4.4 Firefly Synchronization

Fireflies provide one of the most spectacular examples of synchronization in Nature. Consider a tree, 45 feet tall, having firefly on each of the leaf and all fireflies flash in perfect coordination, at a specific rate. This phenomenon is observed at night in certain parts of the world. This light pattern is a part of their mating display. Each species of firefly has a characteristic flash pattern that helps the male and female recognize each other. Male species fly and flash and females respond with flash. The production of light is called bioluminescence. It involves highly efficient chemical reactions that result in release of particles of light with no emission of heat. Fireflies combine chemical luciferin and oxygen with enzyme luciferase in their lanterns (part of their abdomen) to produce light (NPS 2012).

Wakamiya and Murata apply a model of pulse coded oscillators to data gathering in WSN (Tyrelle 2006). The purpose of using pulse coded oscillator is that each sensor node independently determines the cycle and timing at which it emits a message. For efficient data gathering, it has an effect that sensor information is propagated in concentric circles from edge of network to base station. It means that all sensor nodes equidistant from the base station must transmit their information simultaneously, but slightly before their neighboring nodes. When the sensor node receives a radio signal with a level smaller than it's own, it is stimulated and it raises its state as seen from the equation.

$$x_i(t^+) = B(x_i(t) + \in)$$ (4.7)

where,

$$B(x) = \begin{cases} 0 & \text{if } x < 0 \\ 1 & \text{if } x > 1 \\ x & \text{otherwise} \end{cases}$$

At the desired frequency of data gathering, base station periodically broadcasts beacon signals within its RF transmission range. Sensor nodes which receive this signal recognize that they are within the innermost of concentric circles and set their level as 1. In addition, by repeatedly being stimulated by beacon signals, they become synchronized.

4.5 Bacterial Foraging Optimization

Bacterial foraging optimization is the novel procedure proposed by Passino (2002) used for optimization and control. Microscopic organisms such as bacteria scan for supplements in a way to augment vitality obtained per unit time. Single bacterium additionally speaks with others by sending signs. A bacterium takes scavenging choices in the wake of considering two past variables. The procedure in which a bacterium moves by making little strides while hunting down supplements is called chemotaxis. The key thought of bacteria foraging is emulating chemotactic development of virtual microbes in the problem search space. The optimization technique used by them can be divided into following steps:

- *Chemotaxis*: A bacterium regularly pulls itself to move to the closest sources of supplements. The procedure is the primary operation that speaks to the development of a bacterium accomplished through two interchangeable ways such as *swimming* and *tumbling* by flagella. Swimming is for the most part performed in one heading either upwards or downwards and tumbling is performed in clockwise or anticlockwise savvy course to achieve supplement and maintain a strategic distance from poisonous environment way.

  Now, suppose $\theta^i(j, k, l)$ represents the ith bacterium at jth chemotactic, kth reproduction, and lth elimination and dispersal steps, and $C(i)$ is the random length unit. Then in computational chemotaxis the movement of the bacterium may be represented by

$$\theta^i(j+1,\,k,\,l) = \theta(j,\,k,\,l)C(i)\frac{\Delta(i)}{\sqrt{\Delta^i(i)\Delta(i)}} \tag{4.8}$$

where, $\Delta(i)$, Δ is random vectors on $[-1,\,1]$.

$\frac{\Delta(i)}{\sqrt{\Delta^i(i)\Delta(i)}}$ is the unit walk in the random direction C: Run length unit.

$\theta^i(j+1,\,k,\,l)$ and $\theta^i(j,\,k,\,l)$ represents the location of the bacterium.

- *Swarming*: When a gathering of *E. coli* microorganisms is situated in the semisolid agar having a sensor to an unsafe spot, they move from the middle to outwards course in a ring structure of microscopic organisms by following the supplement inclination delivered by the gathering of microbes which devour the supplement. The spatial request relies upon both the outward developments of the ring and the neighborhood arrivals of the attractant, which works as a fascination sign between microbes to assemble into a swarm. The cell-to-cell signaling in bacterium can be represented by the following function:

$$
\begin{aligned}
J_{CC}(\theta, P(j,\,k,\,l)) &= \sum_{i=1}^{S} J_{CC}\left(\theta, \theta^i(j,\,k,\,l)\right) \\
&= \sum_{i=1}^{S}\left[-d_{\text{attractant}}\exp\left(-w_{\text{attractant}}\sum_{m=1}^{p}(\theta_m - \theta_m^i)^2\right)\right] \\
&\quad + \sum_{i=1}^{S}\left[-h_{\text{repellant}}\exp\left(-w_{\text{repellant}}\sum_{m=1}^{p}(\theta_m - \theta_m^i)^2\right)\right] \tag{4.9}
\end{aligned}
$$

where,

$J_{CC}(\theta, P(j,k,l))$ is the objective function value to be added to the actual objective function (to be minimized) to present a time-varying objective function,

S is the total number of bacteria,

p is the number of variables to be optimized, which are present in each bacterium,

$\theta = [\theta_1 \theta_2 \ldots \theta_p]^T$ is a point in the p-dimensional search domain, $-d_{\text{attractant}}$, $-w_{\text{attractant}}$, $-h_{\text{repellant}}$, $-w_{\text{repellant}}$ are different coefficients that should be chosen properly.

- *Reproduction*: The weakest solid microorganism, i.e., bacteria in the long run dies eventually while each of the more beneficial microbes (those yielding lower estimation of the goal capacity) abiogenetically split into two microscopic organisms, which are then put in the same area. This keeps the swarm size consistent. The health is computed by the following equation:

$$J^i_{\text{health}} = \sum_{j=1}^{N_c+1} J(i, j, k, l) \qquad (4.10)$$

where,

N_c is the number of chemotactic steps and
j is the error value

- *Elimination and Dispersal*: Gradual or sudden changes in the nearby environment where a bacterium population lives may happen because of different reasons, e.g., a noteworthy nearby ascent of temperature may execute a gathering of microscopic organisms that are presently in a locale with a high grouping of supplement inclinations. Occasions can occur in such a style to the point that all the microscopic organisms in a locale are executed or a gathering is scattered into another area. To reenact this wonder in bacterial rummaging, some microorganisms are liquidated indiscriminately with a little likelihood while the new substitutions are haphazardly instated over the search space.

The complete algorithm for the same is as explained below in Algorithm 4.5 (Das et al. 2009a, b):

Algorithm 4.5: Bacterial Foraging Optimization Algorithm

Input: Parameters

   p dimension of the search space,
   S the number of bacteria in the colony,
   N_c chemotactic steps
   N_s swim steps
   N_{re} reproductive steps
   N_{ed} eliminiation and dispersal steps
   P_{ed} Probability of eliminiation
   $C(i)(i=1,2,....S)$ run-length unit (the size of step taken in each run or tumble)
   θ_i

Output: Set of bacterium

1 elimination and dispersal: l=l+1
2 reproduction loop: k=k+1
3 chemotaxis loop: j=j +1
4   For i =1,2,..., S , take a chemotactic step for bacterium i as follows:
5   Compute fitness function, J (i, j, k, l)
        Let $J(i,j,k,l) = J(i,j,k,l) + J_{cc}(\theta^i(j,k,l),P(j,k,l))$
        (i.e. add on to cell to cell attractant repellant profile to simulate the swarming)
6   Let J_{last}=J (i, j, k, l) to save this value since we may find better value via a run.
7   Tumble: Generate a random vector $\Delta(i) \in R^n$ with each element $\Delta_m(i)$ m = 1,2,..., p, a random number on [1, -1]
8   Move: C(i) in the direction of the tumble for bacterium i
        $$\theta^i(j+1,k,l) = \theta(j,k,l)C(i)\frac{\Delta(i)}{\sqrt{\Delta^i(i)\Delta(i)}}$$
9   Compute J (i, j+1, k, l) and let
        $$J(i,j+1,k,l) = J(i,j,k,l) + J_{cc}(\theta^i(j,k,l),P(j,k,l))$$

---

10   Swim:
                Let m=0 (counter for swim length)
                While m<N$_s$
                    If J(i,j+1,k,l)<J$_{last}$
                        J$_{last}$=J(i, j+1, k , l)
                        $$\theta^i(j+1,k,l) = \theta(j,k,l)C(i)\frac{\Delta(i)}{\sqrt{\Delta^T(i)\Delta(i)}}$$
                        Use $\theta^i(j+1,k,l)$ to compute new J(i, j+1, k, l)
                    Else
                            m=N$_s$
                    Go to next bacterium (i+1) if i≠ S (Step 5)
11 If j<N$_c$, Go to step 3
12 Reproduction
        For given k and l, and for each i=1,2....,S
                $J^i_{health} = \sum_{j=1}^{N_c+1} J(i,j,k,l)$ be the health of bacterium i
                Sort bacteria and chemotactic parameters C(i) in order of
                ascending cost J$_{health}$ (Higher cost implies lower health)
            S$_r$ bacteria with the highest J$_{health}$ dies and remaining S$_r$ bacteria with the
            best values split
13 if k<N$_{re}$, Go to step 2
14 Elimination dispersal:
        For i=1,2....,S with probability P$_{ed}$ eliminate and disperse each bacterium
        if a bacterium is eliminated, simply disperse another one to a random
        location on the optimization domain
        if l<N$_{ed}$
            Step 1
        Else
            End

Bacterial foraging optimization technique is used for variety of optimization problems in networks (Kim et al. 2007; Das et al. 2009a, b; Shen et al. 2009). For instance, this algorithm has been chosen and connected in feedforward neural system to improve the learning procedure as far as convergence rate and classification precision (Al-Hadi et al. 2011). One of the fundamental procedures in bacterial foraging optimization calculation is the chemotactic development of a virtual bacterium that makes a trial arrangement of the optimization. This procedure of chemotactic development is guided to make the learning procedure of artificial neural network quicker. The created bacterial foraging optimization algorithm feedforward neural network is thought about against particle swarm optimization feedforward neural network. The outcomes demonstrate that bacterial foraging optimization algorithm gave a superior execution as far as convergence rate and classification precision contrasted with particle swarm optimization feedforward neural network. Nevertheless, it has also been used to solve the congestion control issue in Abharian and Shakeri (2011). Taking into account the bacterial foraging optimization technique, a straightforward and successful vigorous bacterial foraging random early detection (BF-RED) has been proposed. RED parameter can be specifically acquired by tackling the predetermined enhancement issue by means of foraging optimization technique. The simulation results uncover that the proposed plan is better than the current active queue management (AQM) plans. Results obtained from simulations in this work demonstrated that higher goodput per association were accomplished when the strong AQM controller was utilized when contrasted with RED. In addition, lower quantities of retransmission timeout and RTT qualities were acquired when vigorous AQM was utilized. These outcomes show that vigorous AQM is a potential contender for selection as a customary AQM approach in systems that utilize a TCP convention. Furthermore, bacteria foraging optimization is also used in solving routing problem in wireless sensor networks (WSN) (Chen et al. 2012; Kulkarni and Venayagamoorthy 2011).

4.6   Cuckoo Search

Cuckoo search is another recent and promising algorithm used as an optimization technique proposed by Yang and Deb (2009). Cuckoos are intriguing fowls not just on account of the wonderful sounds they can make, but also as a result of their forceful proliferation technique. A few cuckoos' animal categories, for example, the *ani* and *Guira* lay their eggs in mutual homes; however, they may evacuate others' eggs to expand the likelihood of their own eggs (Payne et al. 2005). It depends on three admired guidelines:

1. Every cuckoo lays one egg at once, and dumps its egg in an arbitrarily picked home;
2. The best homes with high caliber of eggs will extend to the cutting edge;
3. The quantity of accessible hosts homes is same and the egg laid by a cuckoo is found by the host winged creature with a likelihood $p_a \in (0, 1)$. Finding operation on some arrangement of the most exceedingly bad homes and dumping arrangements from more remote estimations is used for discovery.

Some host winged animals can draw in direct clash with the intruding cuckoos. On the off chance that a host winged creature finds the eggs are not their claims, they will either discard these outsider eggs or essentially relinquish its home and construct another home somewhere else. Some cuckoo species, for example, the New World brood-parasitic *Tapera* have advanced in a manner that female parasitic cuckoo are regularly spending their significant time in the mimicry and shading of the eggs of a couple picked host animal types. These decrease the likelihood of their eggs being relinquished and hence build their reproduction. Moreover, the timing of egg-laying of a few animal types is additionally stunning.

Different studies have demonstrated that flight conduct of numerous creatures and creepy crawlies has exhibited the commonplace attributes of Levy flights. The work presented by Yang and Deb (2009) uses cuckoo search via Levy's flight phenomena using the following Algorithm 4.6 (Yang and Deb 2009):

Algorithm 4.6: Cuckoo Search via Levy Flight

$Input$: Optimization function $f(x), x = (x_1, ... x_d)^T$
            Initial population of n host nests x_i (i=1,2,.....n)
$Output$: set of cuckoo
1   While (t< MaxGeneration) or (stop criterion)
2   Pick a random cuckoo by levy flight and evaluate its quality/fitness F_i
3   Choose a nest among n (say, j) randomly
4   If ($F_i > F_j$)
5   Replace j by the new solution
6   End
7   A fraction (p_a) of worse nests are abandoned and new ones are built
8   Keep the best solutions
9   Rank the solution and find the current best
10   End while
11   Post process results and visualization

Applications of cuckoo search are used in different network optimization issues (Yeng and Deb 2010, 2014; Valian et al. 2013). An efficient computation approach based on cuckoo search has been proposed for data fusion in wireless sensor networks (Dhivya et al. 2011; Dhivya and Sundarambal 2011). Additionally, it is being used to solve the feedforward neural network classification problem (Valian et al. 2011). Nevertheless, cuckoo search is adapted to solve NP-hard combinatorial optimization problems like traveling salesman problem (Ouaarab et al. 2014). A new quantum-inspired cuckoo search was developed to solve Knapsack problems, which shows its effectiveness (Layeb 2011). A conceptual comparison of the cuckoo search, particle swarm optimization, differential evolution, and artificial bee colony algorithms is shown in Civicioglu and Besdok (2013). The work presented by them shows cuckoo search proves to be a very effective model by providing with robust and precise results in comparison to other models.

4.7   Other Inspirations

Prominent problems in networks such as routing, streaming, coding, and connectivity can be solved by learning from social insects. This chapter explained the process by which the network systems can be biologically inspired in solving some of the major problems. Interestingly, bee colony can also be used for routing purposes (Wedde et al. 2004, 2005). An emerging category of social insects like gray wolves, bats, and fish can be used in solving various other problems as well (Yang 2010; Mirjalili et al. 2014). Swarm robotics is another application of using swarm principles to robots. Fish schooling can be used for adaptive flocking and tracking enabling a swarm of autonomous mobile robots to navigate toward achieving a mission (Lee and Chong 2008).

4.8 Summary

Social insects are, in a way, beautiful crawlers that perform their task in a contrasting manner by interacting with their own species. The task is species dependent whether it is flying in a synchronized manner in sky, or reproducing for the next generation, or performing foraging. The present chapter gives details on species such as ants, bees, fireflies, cuckoo, bacteria, and birds and how they are used in solving networking issues such as routing, Huffman coding, synchronization in wireless sensor networks, etc.

References

Abharian, A., & Shakeri, E. (2011). Bacteria foraging optimization Robust-RED for AQM/TCP network. *International Journal of Modeling and Optimization, 1*(1), 49.

Al-Hadi, I. A. A., Hashim, S. Z. M., & Shamsuddin, S. M. H. (2011). Bacterial foraging optimization algorithm for neural network learning enhancement. In *11th International conference on hybrid intelligent systems (HIS)* (pp. 200–205).

Baguda, Y. S., Fisal, N., Rashid, R. A., Yusof, S. K., Syed, S. H., & Shuaibu, D. S. (2012). Biologically-inspired optimal video streaming over unpredictable wireless channel. *International Journal of Future Generation Communication and Networking.*

Beckers, R., Holland, O. E., & Deneubourg, J. L. (2000). From local actions to global tasks: Stigmergy and collective robotics. In *Studies in cognitive systems* (Vol. 26, pp. 1008–1022).

Caro, G. D., & Dorigo, M. (1997). AntNet: A mobile agents approach to adaptive routing. Technical report 97–12, IRIDIA, Universite' Libre de Bruxelles.

Caro, G. D., Ducatelle, F., & Gambardella, L. M. (2005). AntHocNet: An adaptive nature-inspired algorithm for routing in mobile ad hoc networks. *European Transactions on Telecommunications, 16,* 443–455.

Chen, Z., Li, S., Yue, W., Hu, L., & Sun, W. (2012, November). Bacterial foraging optimization algorithm based routing strategy for wireless sensor networks. *International Review on Computers and Software, 7*(6), 2826–2830.

Ciimara, D., & Loureiro, A. A. F. (2001). A GPS/Ant-Like routing algorithm for ad hoc networks. *Telecommunication Systems, 18*(1–3), 85–100.

Civicioglu, P., & Besdok, E. (2013). A conceptual comparison of the Cuckoo-search, particle swarm optimization, differential evolution and artificial bee colony algorithms. *Artificial Intelligence Review, 39*(4), 315–346.

Das, S., Biswas, A., Dasgupta, S., & Abraham, A. (2009a). Bacterial foraging optimization algorithm: Theoretical foundations, analysis, and applications. *Foundations of Computational Intelligence* (Vol. 3, pp. 23–55). Berlin Heidelberg: Springer.

Das, S., Dasgupta, S., Biswas, A., Abraham, A., & Konar, A. (2009b). On stability of the chemotactic dynamics in bacterial-foraging optimization algorithm. *IEEE Transactions on Systems, Man and Cybernetics, Part A: Systems and Humans, 39*(3), 670–679.

Dhivya, M., Sundarambal, M., & Anand, L. N. (2011). Energy efficient computation of data fusion in wireless sensor networks using cuckoo based particle approach (CBPA). *International Journal of Communications, Network and System Sciences, 4*(4), 249.

Dhivya, M., & Sundarambal, M. (2011). Cuckoo search for data gathering in wireless sensor networks. *International Journal of Mobile Communications, 9*(6), 642–656.

Dorigo, M., & Gambardella, L. M. (1997). Ant colony system: A cooperative learning approach to the traveling salesman problem. *IEEE Transactions on Evolutionary Computation, 1*(1), 53–66.

Great Smoky Mountains. (2012). Retrieved January 2012, from http://www.nps.gov/grsm/naturescience/fireflies.htm.

Gunes, M., Sorges, U., & Bouazizi, I. (2002). ARA-the ant-colony based routing algorithm for MANETs. In *Proceedings of the International Conference on Parallel Processing Workshops* (pp. 79–85).

Kim, D. H., Abraham, A., & Cho, J. H. (2007). A hybrid genetic algorithm and bacterial foraging approach for global optimization. *Information Sciences, 177*(18), 3918–3937.

Kulkarni, R. V., & Venayagamoorthy, G. K. (2011). Particle swarm optimization in wireless-sensor networks: A brief survey. *IEEE Transactions on Systems, Man, and Cybernetics, Part C: Applications and Reviews, 41*(2), 262–267.

Layeb, A. (2011). A novel quantum inspired cuckoo search for knapsack problems. *International Journal of Bio-Inspired Computation, 3*(5), 297–305.

Lee, G., & Chong, N. Y. (2008). Flocking controls for swarms of mobile robots inspired by fish schools. In *Recent Advances in Multi Robot Systems*.

Leibnitz, K., Wakamiya, N., & Murata, M. (2006). Biologically inspired self-adaptive multipath routing in overlay networks. *Communications of the ACM, 49*(3), 63–67.

Mirjalili, S., Mirjalili, S. M., & Lewis, A. (2014). Grey Wolf optimizer. *Advances in Engineering Software, 69*, 46–61.

Nicole, A. (2012). Fish, networks, and synchronization. Ph.D. (Vol. 199, pp. 3518–3562).

Ouaarab, A., Ahiod, B., & Yang, X. S. (2014). Discrete cuckoo search algorithm for the travelling salesman problem. *Neural Computing and Applications, 24*(7–8), 1659–1669.

Passino, K. M. (2002). Biomimicry of bacterial foraging for distributed optimization and control. *IEEE Control Systems, 22*(3), 52–67.

Payne, R. B., Sorenon, M. D., & Klitz, K. (2005). The cuckoos. Oxford: Oxford University Press.

Perretto, M., & Lopes, H. S. (2005). Reconstruction of phylogenetic trees using the ant colony optimization paradigm. *Genetics and Molecular Research, 4*(3), 581–589.

Shen, H., Zhu, Y., Zhou, X., Guo, H., & Chang, C. (2009). Bacterial foraging optimization algorithm with particle swarm optimization strategy for global numerical optimization. In *Proceedings of the First ACM/SIGEVO Summit on Genetic and Evolutionary Computation* (pp. 497–504). New York: ACM.

Tatar, N., & Holban, S. (2012). *A Bio Inspired Alternative to Huffman Coding*. Suceava, Romania: Proceeding of Development and Application Systems.

Tyrrell, A., Auer, G., & Bettstetter, C. (2006). Fireflies as role models for synchronization in ad hoc networks. In *Proceedings of the International Conference on Bio Inspired Models of Network, Information and Computing Systems* (pp. 4.

Valian, E., Mohanna, S., & Tavakoli, S. (2011). Improved cuckoo search algorithm for feedforward neural network training. *International Journal of Artificial Intelligence and Applications, 2*(3), 36–43.

Valian, E., Tavakoli, S., Mohanna, S., & Haghi, A. (2013). Improved cuckoo search for reliability optimization problems. *Computers and Industrial Engineering, 64*(1), 459–468.

Wedde, H. F., Farooq, M., & Zhang, Y. (2004). Beehive: An efficient fault- tolerant routing algorithm inspired by honey bee behavior. *Ant Colony, Optimization, and Swarm Intelligence, 3172*, 8394.

Wedde, H.F. et al. (2005). BeeAdHoc: An energy efficient routing algorithm for mobile ad-hoc networks inspired by bee behavior. In *Proceedings of the GECCO* (pp.153–160). New York: ACM.

Yang, X. S. (2010). A new metaheuristic bat-inspired algorithm. In *Nature Inspired Cooperative Strategies for Optimization*. Studies in Computational Intelligence (pp. 65–74).

Yang, X. S., & Deb, S. (2009). Cuckoo search via Lévy flights. In *World Congress on Nature and Biologically Inspired Computing, NaBIC* (pp. 210–214).

Yang, X. S., & Deb, S. (2010). Engineering optimisation by cuckoo search. *International Journal of Mathematical Modelling and Numerical Optimisation, 1*(4), 330–343.

Yang, X. S., & Deb, S. (2014). Cuckoo search: Recent advances and applications. *Neural Computing and Applications, 24*(1), 169–174.

Chapter 5
Immunology and Immune System

Abstract Biological immune systems have immense potential in fighting with foreign bodies that encounter the body. It has two protective layers, viz, innate layer and adaptive layer, which helps in dealing with the different types of attacks where the innate layer consists of skin, mucus, and tears and adaptive layer comprises of B-cells and T-cells. Additionally, it has the interesting characteristic of remembering the foreign bodies that attack the human body. This feature helps them in fighting with the foreign bodies at a faster rate. This chapter throws light on the human immune system. Additionally, it explains how the system can help in dealing with different problems of computer networks.

5.1   Human Immune System

Biological immune systems have intelligent capabilities of detecting antigens (foreign bodies in the system) in the body. As shown in Fig. 5.1, the immune system can be classified into two types, innate and adaptive.

Innate immunity is the first line of defense for pathogens. It is nonspecific and is meant for rapid detection and elimination of pathogens. It generally refers to nonspecific defense mechanisms that come into play within hours of an antigen's appearance in the body. It is referred to as nonspecific defense mechanism since it is not designed for any specific pathogen. It can further be classified as physical barriers and blood-borne. Physical barriers such as skin, tears, saliva, and mucus stop the infection before entering the body (Greensmith et al. 2010). If pathogens manage to get past the physical barriers, blood-borne body cells come into the picture. Their response will also be nonspecific. This process, called phagocytosis, is carried out by a number of different phagocytes, the most common types being the neutrophils and macrophages. Neutrophil has molecules on their cell walls that help in identifying foreign particles. Once the foreign particles are identified, it will attach to the microorganism wall, to engulf and enclose the microorganism in the

H. Rathore, *Mapping Biological Systems to Network Systems*,
DOI 10.1007/978-3-319-29782-8_5

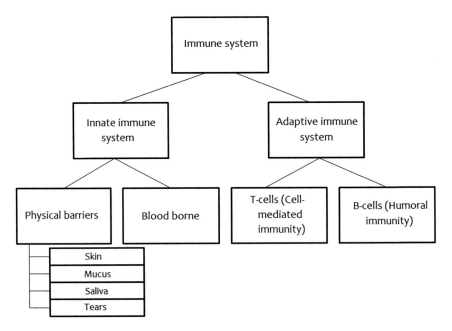

Fig. 5.1 Classification of immune system

Fig. 5.2 Engulfing of
pathogen in neutrophil

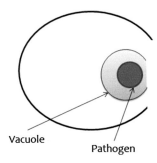

vacuole as shown in Fig. 5.2. Enzymes present in Golgi body in the neutrophil are secreted, which then digest the microorganisms. Macrophages perform the same task outside the blood vessels, so that the pathogens can be removed from tissue fluid.

If the innate immune system cannot remove the pathogen, then the adaptive immune system takes over. Adaptive immune system comprises of a network of cells, tissues, and organs that work in cohesion to protect the body. The cells involved are white blood cells, or leukocytes, which come in two basic types, phagocytes and lymphocytes (Durani 2012). Classification of Leukocytes is depicted in Fig. 5.3. Phagocytes have already been discussed in innate immune system. Lymphocytes are of two types, namely T-cells and B-cells. Leukocytes are

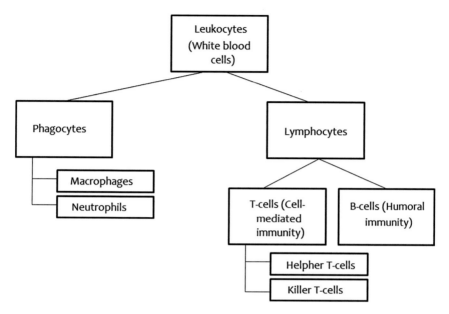

Fig. 5.3 Classification of leukocytes

developed from undifferentiated stem cells in the bone marrow. Lymphocytes start out in the bone marrow, stay there, and mature into B-cells. Alternatively, they leave for the thymus gland, where they mature into T-cells. They are called T-cells because the latter stages of their development occur in the thymus. Spleen, bone marrow, and thymus are also called as lymphoid tissues. Lymph nodes are specialized tissue harboring cells of the innate immune system called leukocytes and macrophages, in addition to specialized cells of the adaptive immune system; T and B cells. These nodes are connected by the lymphatic circulation of the body and help the two arms of human immune system coordinately fight a pathogenic attack in coordination.

The adaptive immune system consists of two complementary systems, namely cellular immune system and humoral immune system. The humoral immune system is aimed at bacterial infections and extracellular viruses, but can also respond to individual foreign proteins. This system contains soluble proteins called antibodies which bind the bacteria, viruses, or large molecules identified as foreign and target them for destruction. Antibodies are produced by B-cells. Antigens are secreted by the pathogens which cause the immune system to respond. B-cells produce and secrete antibodies after they encounter antigens. The cellular immune system destroys host cells infected by viruses and also destroys some parasites. The agents at the heart of this system are a class of T-cells. B-cells act as the body's military intelligence system which alerts the body's defenses to seek and destroy the enemy targets. T-cells act as the soldiers, destroying the invaders that the intelligence system has identified. T-cells are broadly classified into two types, namely helper T-cells and killer T-cells. Killer T-cells interact with infected host cells through receptors on T-cell surface. Helper T-cells interact with macrophages and secrete cytokines that stimulate killer T-cells, helper T-cells, and B-cells to proliferate and produce antibodies specific to the pathogen.

In human immune system, B-cells produce an antibody on encountering with an antigen. As there can be any type of antigen that can excite human immune system, there is a requirement of a specific antibody that can bind to that antigen. In human bodies, B-cells produce n different types of antibodies that bind to the particular antigen to decrease its effect in the body. Further, B-cells are capable of maintaining a memory which keeps track of all the previous invaders. B-cells simulation can be modeled on three parameters:

(1) Affinity between B-cell and pathogen.
(2) Affinity among its neighbors.
(3) Separation (enmity) factor from the loosely connected neighbors of the B-cells.

The general structure of all antibodies is similar. However, a small region at the tip of the protein is extremely variable which allows millions of antibodies to have minute differences in the tip structures, or antigen binding sites, to exist. Typical structure of antibody is as shown in Fig. 5.4. It is composed of 2 chains, one light chain (black color) and one heavy chain (red color). Both the chains consist of constant region and variable region as denoted by C and V respectively. Antibodies are diverse in nature because of the various processes such as domain variability, V (D) J recombination, somatic hyper mutation and affinity maturation, class switching or affinity designations (Wikipedia 2012).

Fig. 5.4 Antibody structure

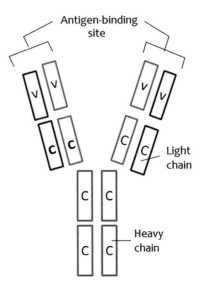

Once the antibodies are produced, they continue to exist in a person's body. It provides better immunity to the living beings in a way that if the same antigen is presented to the immune system again, the antibodies are already there to perform their task. Hence if someone gets sick with a certain disease, like chickenpox, the person typically does not get sick from the disease again. Although antibodies can recognize an antigen and lock onto it, they are not capable of destroying it without the help of other cells (Morphosys 2012). That is the job of the T-cells, which are a part of the system that destroys the antigens which have been tagged by antibodies or cells that have been infected or somehow changed. Furthermore, T-cells helps signal other cells (like phagocytes) to do their tasks. All of these specialized cells and parts of the immune system offer body protection against a disease. This protection is called immunity.

To wrap up, antigens are secreted by the pathogens which cause the adaptive immune system to respond. B-cells produce and secrete antibodies after they encounter antigens. Once they produce a specific antibody for the antigen it forms a complex called antigen-antibody complex which in turn is engulfed by the T-cells. After the B-cells produce antibodies they give rise to plasma cells from which further antibodies are produced for that specific antigen.

Thus, the human immune system has the following aims (Hao 2005):

(1) *Accountability*: Human body tries to identify, find, and destroy the nonself particles that have entered the human body.
(2) *Disposability*: The human body continues to work even if the nonself particles are detected.
(3) *Correction*: The body prevents from attacking its own cells.
(4) *Integrity*: Genetic codes inside the cells are not modified.

5.2 Primary Versus Secondary Response

Primary response in the immune system is defined as the first exposure of the immune system to the antigen. Secondary response occurs when there is another exposure to the same antigen. When the immune system encounters the antigen for the first time, B-cells multiply to form two types of cells, i.e., plasma cells and memory B-cells.

The plasma cells produce antibodies and memory B-cells just keep a track of the type of antigen. Secondary response occurs when another exposure to the same antigen causes the memory B-cells to rapidly form the plasma cells to produce antibodies out of it as shown in Fig. 5.5a. Secondary response is faster since it produces more antibodies than in the case of primary response as shown in Fig. 5.5b. It also has a high magnitude which rapidly decreases the count of antigen rapidly.

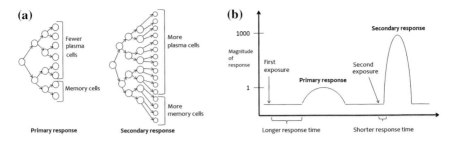

Fig. 5.5 Primary versus secondary response: **a** Clearly it is seen that number of plasma cells and memory cells are high in secondary response as compared to the primary response thereby increasing the number of antibodies; **b** The response time in primary exposure is less compared to the secondary response (*Source* Sharp 2011)

5.3   Artificial Immune Systems

Human beings have its own particular differing and complex interruption recognition framework with the uncommon abilities of learning, memory and adjustment. Artificial Immune Systems (AIS) roused by the natural immune systems are a rising sort of soft computing methods (De Castro and Timmis 2002a, b, c). With the recognizing components of pattern recognition, anomaly detection (Hunt et al. (1996, 1998)), data analysis, and machine learning, the AIS have as of late increased extensive exploration enthusiasm from diverse groups (Dasgupta and Attoh-Okine 1997). Their fruitful industry applications incorporate anomaly detection, optimization, fault diagnosis, pattern recognition, etc. (Dasgupta et al. 2003). Typically artificial immune systems applications lies in data clustering, optimization and fault diagnosis inspired from different phenomenon of human immune systems as shown in following Fig. 5.6 (Gao 2006).

5.3.1   Negative Selection

Artificial immune systems for computers were originally conceived as a defense mechanism for individual hosts against computer viruses. Jeffrey Kephart of IBM

Fig. 5.6 Artificial immune system and its applications

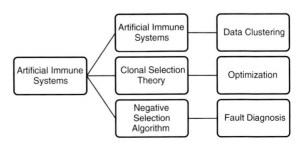

Research and Forrest proposed one of the earliest designs (Kephart 1994; Forrest et al. 1994). While Kephart introduced the concepts of integrity checking, activity monitor and decoy programs, Forrest introduced the concepts of negative selection. Integrity checking was made to check nonself files and programs and accordingly activity monitoring was made to react to the suspicious activity. Decoy programs were created to attract the infection. Integrity checking and activity monitors served as innate immune system and decoy programs served as adaptive immune system.

The human immune system contains an organ called thymus that is situated behind the breastbone, which performs a vital part in the development of T-cells. After T-cells are produced, they relocate into the thymus where they develop. Amid this development, all T-cells that perceive self-antigens are prohibited from the population of T-cells; a procedure termed negative selection. Negative selection comes into picture when nonself cells act as self-cells by mimicking the features of self-cells. Forrest introduced the concept of negative selection by generating a fixed-size repertoire of random strings, deleting any strings that occur in the data that the system is told to protect. When asked to verify data, the integrity checker looks for sub-strings that match the strings in the repertoire. Since all "self" strings were removed from the repertoire, any match implies a verification failure. Furthermore, negative selection algorithm has been proposed in the literature with applications focused on the problem of anomaly detection, such as computer and network intrusion detection, time series prediction, image inspection and segmentation, and hardware fault tolerance (De Castro and Timmis 2002a, b, c).

Pattern recognition is another branch that can be solved by negative selection algorithm. For this, given a suitable problem representation, we characterize the set of patterns to be secured and call it the self-set (P). Based upon the negative selection algorithm, we produce a set of detectors (M) that will be dependable to distinguish all components that don't have a place with the self-set, i.e., the nonself components. The negative selection procedure runs in following manner (Fig. 5.7a):

- Generate arbitrary candidate components (C) utilizing the same representation received;
- Compare (match) the components in C with the components in P. In the event if the match happens, i.e., if a component of P is perceived by a component of C, then discard this component of C; else store this component of C in the detector set M.

After generating the set of detectors (M), the following phase of the calculation constitutes observing the framework for the presence of nonself patterns (Fig. 5.7b). For this situation, expect a set $P*$ of patterns to be ensured. This set may be made out of the set P in addition to other new examples, or it can be a totally novel set. For all components of the detector set, that compares to the nonself examples, check if it perceives (matches) a component of $P*$ and, if yes, then a nonself example was perceived and a move must be made. The subsequent activity of identifying nonself fluctuates as varies accordingly by the issue under assessment and extrapolates the pattern recognition.

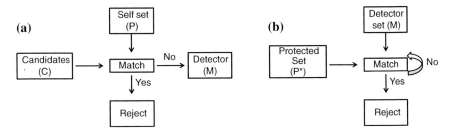

Fig. 5.7 Pattern recognition through the Negative Selection Algorithm **a** Generating set of detectors **b** Monitoring for the presence of nonself patterns (*Source* De Castro and Timmis 2002a, b, c)

5.3.2 Clonal Selection

Clonal selection is the characteristic of human immune system, to respond in front of certain antigen which attacks our body. B–cells and T-cells initially, produce 'n' number of child B-cells and T-cells each having the capability to respond to different types of bacteria seen in the life cycle of human body. Typically B-cells and T-cells are present in the lymphatic vessels of our body to respond when the any foreign body such as virus and bacteria occurs in the body. When the bacteria are present in our body, the dendritic cell responds as Antigen Presenting Cells (APC) for the T-cells to react to it as shown in Fig. 5.8. B-cells responding to that particular type of bacteria directly react to it which enters our body. These B-cells

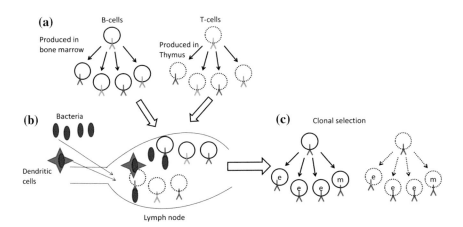

Fig. 5.8 *Clonal Selection Process*: **a** B-cells and T-cells produce different child B-cells and child T-cells to bind to different types of bacteria (represented by different color). **b** Later, when bacteria are seen in the vicinity, dendritic cells take these bacteria in the lymphatic nodes for the T-cells to respond to it (shown in *red* color). B-cells directly react to bacteria. **c** Finally, more B-cells and T-cells are produced in the body in the form of effector cells **e** and memory cells (m) (shown in *red* color)

and T-cells which encounters the bacteria creates clones of them to respond to the rest of the bacteria present in the body. The overall process is called clonal selection process.

The work introduced by De Castro and Von Zuben (2002) propose a computational execution of the clonal selection principle that unequivocally considers the affinity maturation of the immune response. The general algorithm, named CLONal selection ALGorithm (CLONALG), is primarily derived to perform machine-learning and pattern recognition tasks, and after that it is adjusted to take care of optimization problems, emphasizing multimodal and combinatorial optimization.

Given a set of pattern to be perceived and recognized (P), the fundamental strides of the CLONALG algorithm is shown in Algorithm 5.1.

Algorithm 5.1: CLONALG Algorithm
Input: A set of pattern to be perceived and recognized *(P)*
Output: Recognised pattern
1 Randomly initialize a populace of individual *(M)*;
2 repeat until a sure rule is met, for example, such as a minimum pattern recognition or classification error
3 $\qquad$ For every pattern of *P*, present it to the populace *M* and focus its affinity (match) with every component of the populace *M*;
4 $\qquad$ Select n_1 of the best most affinity elements of *M* and produce duplicates of these individuals relatively to their affinity with the antigen. The higher the affinity, the higher the quantity of duplicates, and the other way around;
5 $\qquad$ Mutate every one of these duplicates with a rate corresponding to their liking with the input pattern: the higher the affinity, the smaller the mutation rate, and the other way around.
6 $\qquad$ Add these mutated individuals to the populace *M* and re-select n_2 of these maturated (optimized) individual to be kept as recollections of the framework.

Optimization Immune Algorithm (opt-1A) is another such model which works on the principle of clonal selection theory (Cutello et al. 2004). A comparative analysis between the two models, viz CLONALG and opt-1A is discussed and addressed on different classes of problems: toy problems (one-counting and trap functions), pattern recognition, numerical optimization problems and NP-complete problem. The work shows opt-1A presents better performance and robustness in comparison to the CLONALG method (Cutello et al. 2005). Optimization problem using clonal selection principle is studied in variety of aspects (Jiao et al. 2005; De Castro and Timmis 2002a, b, c; Cortés and Coello 2003).

5.3.3 Artificial Immune Systems

Later, various features of immune systems were adapted in solving various networking issues. Artificial Immune Systems (AIS) is one such emerging computer science technique which is inspired from this biological process. AIS approach can also be used for data analysis (Timmisa et al. 2000; Timmisa and Neal 2001). An elaborative version of research in AIS domain was given by Dasgupta et al. (2003).

Since the biological immune system have exceptional characteristics and features such as, pattern recognition, feature extraction, diversification, learning capability, inbuilt memory, distributed detection, self regulation etc. By using the ideas from self organization, B-cells proliferation (antigen and antibody), primary and secondary response, hence various networking issues can be solved. AIS are one such branch which uses the machine learning concepts coupled with immune systems. Hunt and Cooke (1995, 1996), Cooke and Hunt (1995) proposed the initial concept of machine learning and immune systems. They created a supervised machine learning mechanism to classify DNA sequences as either promoter or non-promoter classes, by creating a set of antibody strings that could be used as explained in Algorithm 5.2. It is claimed that the classification performed using AIS produces an error rate of 3 %, which outperforms many classification techniques (Timmis and Knight 2001).

Algorithm 5.2: AIS to Recognize Promoter Sequence DNA
Input: Antigen population,
B-cell population
Output: Recognised Promoter Sequence DNA
1 repeat until convergence
2	random antigen selection
3	random point selection in B-cell population to insert into antigen
4	Percent selection of B cells local to the insertion point
5	for each B-cell selected
6	present the antigen to each B-cell to check the antigen antibody binding and check the stimulated B-cell.
7	If B-cell is stimulated enough,
8	clone the B cell
9	otherwise
10	generate a new B-cell, which can bind the antigen
11	Order the B cells by stimulation level
12	Weakest 5% of the population is removed
13	Generation of 25% of new B cells
14	Select 5% B cells to join the immune network

Additionally, AIS is used in unsupervised learning purposes (Timmis et al. 2000). The authors proposed AIS consisting of a set of B cells, antigen training data, links between those B cells. Furthermore, cloning and mutation operations were performed on the B cell objects. Fisher Iris data set (Fisher 1936) was tested to extract meaningful clusters and was visualized in a specially developed tool (Timmis 2001) for exploratory analysis. The authors used these clusters to create a rule set, which in turn is used for classification.

Development of network intrusion system using adaptive immune system has been studied by Hofmeyr and Forrest (2000). They developed a classifier system which contains detector having some contiguous bits as the match rule. Development of such intelligent systems has provided efficient computer security.

Alonso and Nino (2015) modeled a repertoire on AIS as explained in Algorithm 5.3. The model differs from other models in terms of genetic search. The model focuses on regulating the size of each clone based on the interaction between antigen and other clones. It also used log based activation function for the regulation of size of clone.

Algorithm 5.3: Repertoire Completeness
Input: Initialized network
Output: Information Extraction
1 for each antigen
2   do
3   Calculate antigen clone affinities
4   Calculate affinities between clones
5   Calculate activation for each clone
6   Apply activation function
7   Update clone size

Artificial immune systems are also used in data clustering problems. Clustering is useful in several exploratory pattern analyses, grouping, decision-making and machine-learning tasks, including data mining, knowledge discovery, document retrieval, image segmentation and automatic pattern classification. AiNET is one of the advancement of the artificial immune systems in solving data clustering problem. The network evolved is capable of reducing redundancy and describing data structure, including their spatial distribution and cluster interrelations (De Castro and Von Zuben 2001). Resource Limited Artificial Immune System (RLAIS) is another approach towards data analysis proposed by Timmis and Neal (2001). It uses artificial recognition balls (ARBs), which was inspired in describing antigenic interaction within an immune network (Leung et al. 2007). A comparative analysis between different artificial immune models is presented in Galeano et al. (2005).

5.3.4 Pattern Recognition

Pattern recognition, a branch in machine learning is utilized to find similarity of information, i.e., to discover and perceive concealed data from information. Feature extraction and classification are the major tasks in pattern recognition. Immune systems are utilized as a model for pattern recognition and classification (Carter 2000). Pattern recognition and artificial immune systems can be connected in following ways:

- The first perspective to present is the most pertinent representations to be connected to model self and nonself patterns. Here, self-patterns relate to the AIS's segments in charge of perceiving the input pattern (nonself).
- Secondly, the mechanism by which the evaluation of the degree of match (affinity) or degree of recognition, of an input pattern by a component of the AIS is made.

To model immune cells, molecules and the antigenic patterns, the shape space methodology is proposed in Perelson and Oster (1979). The shape-space methodology proposes that an attribute string $s = <s_1, s_2 \ldots s_L>$ in an L-dimensional shape space, S, $(s \in S^L)$ can represent any immune cell or molecule. Each attribute of this string is supposed to represent a feature of the immune cell or molecule. The type of attributes used to represent the string partially to characterize incompletely the shape-space under study is exceptionally reliant on the issue area. Any shape-space developed from a limited letters in order of length k constitutes a k-ary Hamming shape-space. As an illustration, an attribute string based upon the arrangement of binary elements {0, 1} compares to a binary Hamming shape-space (De castro and Von Zuben 1999). It can be considered, for this situation, of an issue of perceiving an arrangement of characters represented by matrices composed of 0's and 1's. Each element of a matrix relates to a pixel in the character.

In the event that the components of s are spoken by real-valued vectors, then we have a Euclidean shape-space. The vast majority of the AIS found in the literature utilize paired Hamming or Euclidean shape-spaces. Different sorts of shape-spaces are likewise conceivable, for example, symbolic shape-spaces, which join diverse (typical) characteristics in the representation of a solitary string s. These are generally found in information mining applications, where the information may contain typical data like age, name of an arrangement of patterns.

The accessibility of an arrangement of N examples (antigens), p_i, $i = 1, \ldots N$ $(p_i \in P)$ to be recognized, and a set of M immune cells and/or molecules (antibodies) $m_j, j = 1, \ldots M$ $(m_j \in M)$ to be used as pattern recognisers (via negative, clonal or network algorithms). Assume additionally, that both have the same length $L(p_i, m_j \in S^L)$. Consider first the binary Hamming shape-space case, which is the most prominently used. Several expressions can be utilized in the degree's determination of match or affinity between a component of P and a component of M. The least complex case is to just ascertain the Hamming separation (DH) between these two components, as given by Eq. (5.1). Another methodology is to hunt down an arrangement of r-adjacent bits (Forrest et al. 1997), and if the quantity of r-coterminous matches between the strings is more prominent than a given limit, then acknowledgment is said to have happened. As the last approach to be mentioned here, we can describe the affinity measure of Hunt (Hunt and Cooke 1997), given by Eq. (5.2). This method has the advantage that it favors sequences of complementary matches, accordingly scanning for comparative areas between the quality strings (patterns) (De Castro and Timmis 2002a, b, c):

$$D_H = \sum_{i=1}^{L} \delta, \text{ where } \delta = \begin{cases} 1 & \text{if } p_i \neq m_i \\ 0 & \text{otherwise} \end{cases} \tag{5.1}$$

$$D = D_H + \sum_i 2^{l_i} \tag{5.2}$$

where, l_i is the length of the ith sequence of matching bits longer than 2.

In the case of Euclidean shape-spaces, the Euclidean distance can be used to determine the affinity between any two components of the system. Other approaches such as the Manhattan distance may also be used.

The applications of artificial immune systems are vast, ranging from machine learning to robotic autonomous navigation. The problem of protecting computers (or networks of computers) from viruses, unauthorized users, etc., constitutes a rich field of research. Spread of viruses in a computer network and spread of malicious information in a social network are two examples of real-world situations that can happen in a complex network (Medline Plus 2012). These two problems are very vital and important in today's networking scenario. It is critical that we derive inspiration from biological immune systems to learn how to protect the networks against such attacks. In the next chapter, these two problems are addressed.

5.4   Summary

Human immune systems have intelligent capability in detecting and eliminating the foreign bodies which attack our system. The chapter discusses the human immune system and some of the recent advances in the artificial immune field. The chapter focus primarily lies in the development of Artificial Immune System for computer security. Many researchers are engaged in producing novel designs for issues in networking systems such as spread of rumor in social networks, epidemic spreading, which are discussed in later chapters.

References

Alonso, O. M., & Nino, L. F. (2015). A New Artificial Immune Network Model based on Repertoire Completeness Assumption.

Carter, J. H. (2000). The immune system as a model for pattern recognition and classification. *Journal of the American Medical Informatics Association, 7*(1), 28–41.

Cooke, D. E., & Hunt, J. E. (1995). Recognising promoter sequences using an artificial immune system. In: *Proceedings of Intelligent Systems in Molecular Biology, 95*, AAAI Press.

Cortés, N. C., & Coello, C. A. C. (2003). Multiobjective optimization using ideas from the clonal selection principle. In *Genetic and Evolutionary Computation—GECCO 2003* (pp. 158–170). Berlin: Springer.

Cutello, V., Nicosia, G., & Pavone, M. (2004). Exploring the capability of immune algorithms: A characterization of hypermutation operators. In *Artificial Immune Systems* (pp. 263–276). Berlin: Springer.

Cutello, V., Narzisi, G., Nicosia, G., & Pavone, M. (2005). Clonal selection algorithms: A comparative case study using effective mutation potentials. In *Artificial Immune Systems* (pp. 13–28). Berlin: Springer.

Dasgupta, D., & Attoh-Okine, N. (1997). Immunity-based systems: A survey. In *1997 IEEE International Conference on Systems, Man, and Cybernetics, 1997. Computational Cybernetics and Simulation* (Vol. 1, pp. 369–374). IEEE.

Dasgupta, D., Ji, Z., & Gonzalez, F. (2003). Artificial immune system (AIS) research in the last five years. In *Proceedings of IEEE Congress on Evolutionary Computation* (Vol. 1, pp. 123–130).

De Castro, L. N., & Timmis, J. (2002a). *Artificial immune systems: A new computational intelligence approach*. London: Springer Science & Business Media.

De Castro, L. N., & Timmis, J. (2002b). Artificial immune systems: a novel paradigm to pattern recognition. *Artificial Neural networks in pattern Recognition, 1*, 67–84.

De Castro, L. N., & Timmis, J. (2002c). An artificial immune network for multimodal function optimization". In *Proceedings of the 2002 Congress on Evolutionary Computation, 2002. CEC'02* (Vol. 1, 699–704). IEEE.

De Castro, L. N., & Von Zuben, F. J. (1999). Artificial immune systems: Part I–basic theory and applications. Technical report, 210Universidade Estadual de Campinas, Dezembro de.

De Castro, L. N., & Von Zuben, F. J. (2001). aiNet: An artificial immune network for data analysis. *Data Mining: A Heuristic Approach, 1*, 231–259.

De Castro, L. N., & Von Zuben, F. J. (2002). Learning and optimization using the clonal selection principle. *IEEE Transactions on Evolutionary Computation, 6*(3), 239–251.

Durani, Y. (2012). Retrieved December 5, 2012, from http://kidshealth.org/parent/general/body_basics/immune.html.

Fisher, R. (1936). The use of multiple measurements in taxonomic problems *Annual Eugenics, 7*(II), 179–188.

Forrest, S., Perelson, A., Allen, L., & Cherukuri, R. (1994). Self-nonself discrimination in a computer. In *Proceedings of IEEE Computer Society Symposium on Research in Security and Privacy* (pp. 202–212).

Forrest, S., Somayaji, A., & Ackley, D. H. (1997). Building diverse computer systems. In *Operating Systems, 1997, The Sixth Workshop on Hot Topics in* (pp. 67–72). IEEE.

Galeano, J. C., Veloza-Suan, A., & González, F. A. (2005). A comparative analysis of artificial immune network models. In *Proceedings of the 7th Annual Conference on Genetic and Evolutionary Computation* (pp. 361–368). ACM.

Gao, X. Z. (2006). Artificial immune systems and their applications. In *NICSO 2006* (p. 7).

Greensmith, J., Whitbrook, A., & Aickelin, U. (2010). Artificial immune systems. In *Handbook of Metaheuristics* (pp. 421–448).

Hao, Y. (2005). Computational intelligence and security. In *International Conference on CIS*.

Hofmeyr, S. A., & Forrest, S. (2000). Architecture for an Artificial Immune System. *Evolutionary Computation, 8*(4), 443–473.

Hunt, J., & Cooke, D. (1995). An adaptive and distributed learning system based on the Immune system. In *Proceedings of IEEE International Conference on Systems Man and Cybernetics (SMC)* (pp. 2494–2499).

Hunt, J., & Cooke, D. (1996). Learning using an artificial immune system. *Journal of Network and Computer Applications: Special Issue on Intelligent Systems : Design and Application., 19*, 189–212.

Hunt, J., King, C., & Cooke, D. (1996). Immunising against fraud. In *Proceedings of Knowledge Discovery and Data Mining and IEE Colloquium* (pp. 38–45). IEEE.

Hunt, J., Timmis, J., Cooke, D., Neal, M., & King, C. (1998). JISYS: Development of an artificial immune system for real world applications. In *Artificial Immune Systems and their Applications* (pp. 157–186). Berlin: Springer.

Jiao, L., Gong, M., Shang, R., Du, H., & Lu, B. (2005). Clonal selection with immune dominance and anergy based multiobjective optimization. In *Evolutionary Multi-Criterion Optimization* (pp. 474–489). Berlin: Springer.

Kephart, J. O. (1994). A biologically inspired immune system for computers. *Proceedings of Artificial Life IV: The Fourth International Workshop on the Synthesis and Simulation of Living Systems* (pp. 130–139). MIT Press.

Leung, K., Cheong, F., & Cheong, C. (2007). Generating compact classifier systems using a simple artificial immune system. *IEEE Transactions on Systems, Man, and Cybernetics, Part B: Cybernetics, 37*(5), 1344–1356.

Medline Plus. (2012). Retrieved December 10, 2012, from http://www.nlm.nih.gov/medlineplus/ency/article/000821.htm.

Morphosys. (2012). Retrieved December 6, 2012, from http://www.morphosys.com/technologies/antibodies/antibody-development-human-body.

Perelson, A. S., & Oster, G. F. (1979). Theoretical studies of clonal selection: minimal antibody repertoire size and reliability of self-non-self discrimination. *Journal of Theoretical Biology, 81*(4), 645–670.

Sharp, D. (2011). Retrieved September 2013, from http://ibbio.pbworks.com/w/page/41532739/Defence%20against%20infectious%20disease%20(HL).

Timmis, J. (2001). aiVIS: Artificial immune network visualization. In *Proceedings of EuroGraphics UK* (pp. 61–69). London: UCL. ISBN 0-9540321-0-1.

Timmis, J., & Knight, T. (2001). Artificial immune systems: Using the immune system as inspiration for data mining. *Data mining: A heuristic approach* (pp. 209–230).

Timmis, J., & Neal, M. (2001). A resource limited artificial immune system for data analysis. *Knowledge-Based Systems, 14*(3—4), 121–130.

Timmis, J., Neal, M., & Hunt, J. (2000). An artificial immune system for data analysis. *Biosystems, 55*(1–3), 143–150.

Wikipedia. (2012). Retrieved December 10, 2012, from http://en.wikipedia.org/wiki/Antibody.

Chapter 6
Information Epidemics and Social Networking

Abstract Communicable disease models have been studied using classical mathematical differential equations for a long time now. It is important to study the communicable disease models, so that one can come up with a good response system to contain the spread of viruses. Social networks are susceptible to the rapid spread of malicious information, commonly referred to as rumors. Rumors often spread rapidly through the network and, if not contained quickly, can be harmful. This chapter describes a method for identifying highly connected nodes in a social network and using these nodes to build immunity against such malicious information. To describe this method, this chapter draws inspiration from two well-established topics in the area of biology: one is the spread of communicable diseases in human population and second is how human body builds immunity against diseases as described in Chap. 5. In case of communicable diseases, it would be very simplistic if we only consider that an infected node can transmit its disease to its nearest neighbors. More realistically speaking, it is possible that an infected node can develop random links with other nodes in the system. The spread of communicable diseases is controlled by both these factors. An infected node with capability to have several random links is capable of spreading the disease through the network faster. We can postulate that certain nodes in a social network exhibit similar behavior and can be defined as highly connected nodes in the network. Once such nodes are identified, the concept of weighting functions is introduced that can be attached to messages passing through such nodes. This chapter describes how the spread of malicious information can be controlled by a community of such highly connected nodes, using the concept of weighted functions.

6.1 Epidemic Spreading

System modeling and analysis of communicable disease models is a very important field of study and has been extensively studied (Goffman and Newill 1964). Such analysis having both of these components would be very helpful in case of pandemic outbreaks such as H1N1 (or swine flu) in 2009 (Jesan et al. 2010). Within a

© Springer International Publishing Switzerland 2016

H. Rathore, *Mapping Biological Systems to Network Systems*,

DOI 10.1007/978-3-319-29782-8_6

short span of 2 months, this disease had rapidly spread across regional boundaries to more than 70 countries, with over 30,000 confirmed cases. H1N1 is one of the many examples of communicable diseases, which threaten us today in this highly connected world. In a diversely distributed country like ours, it is hard to physically measure the spread and impact of such diseases. In such situations, system modeling techniques can prove to be a very effective tool to measure the susceptibility, infection, and recovery rate among people. A very important aspect governing the spread of communicable disease is illustrated by the dynamics cited in Rathore et al. (2012). Here, the authors talk about the impact of time scales on the spread of disease.

Communicable disease spreads in a human population, both through near-neighbor effects and random spatiotemporal links. The spread of an epidemic is described by the math field of Percolation Theory (Hethcote 1976), and can be studied by the numerical analysis of the governing differential equations. In 1927, W.O. Kermack and A.G. McKendrick created a model in which they considered a fixed population with only three compartments, susceptible: $s(t)$, infected: $i(t)$, and recovered: $r(t)$. The compartments used for this model consist of three classes, where

$s(t)$ is used to represent the number of individuals not yet infected with the disease at time t, or those susceptible to the disease.

$i(t)$ denotes the number of individuals who have been infected with the disease at time t and are capable of spreading it to others.

$r(t)$ denotes the number of individuals who have been infected and then recovered from the disease.

$$\frac{ds}{dt} = -k_1 \times s(t) \times i(t) \tag{6.1}$$

$$\frac{di}{dt} = -k_2 \times i(t) + k_1 \times s(t) \times i(t) \tag{6.2}$$

$$\frac{di}{dt} = k_2 \times i(t) \tag{6.3}$$

where k_1 is the infection rate and k_2 is the recovery rate.

Figure 6.1 shows the graph of susceptible, infected, and recovered patients as a function of time, for values of $k_1 = 0.2$ and $k_2 = 0.7$. The figure follows a familiar pattern, where the number of infected patients first increases and once the number of infected patients reaches an equilibrium state, it then starts decreasing at the recovery rate of k_2. Once the number of recovered patients is more than the number of infected patients, the disease will no longer spread in the network. The amount of time from $t = 0$ to the time when the disease stops spreading is referred as information lifetime or time to live (TTL).

It is also instructive, at times, to look at the number of infected and susceptible patients versus the number of recovered patients. This is shown in Fig. 6.2. In this

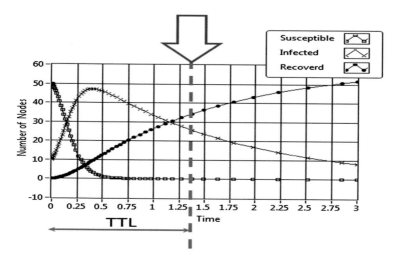

Fig. 6.1   Numerical analysis of SIR model

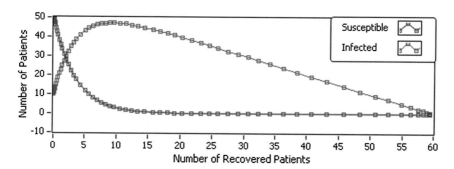

Fig. 6.2   Susceptible and infected versus recovered

figure, the initial conditions are that the number of recovered patients is equal to 0 and that of infected patients is equal to 10. Then, as time progresses, the number of infected patients grows, till it reaches an equilibrium point and then starts decreasing. At the end of the simulation, the number of recovered patients is 60 and the number of infected patients is zero. Likewise, with 60 susceptible patients and 0 susceptible patients, both the infected and susceptible patients approach zero as simulation is complete. In the analysis so far, consideration is only over one specific value of k_1 and k_2. It is often instructive to analyze the results for a ratio of $\frac{k_1}{k_2}$.

Let us denote this ratio by β. Also, there are two observation parameters associated called the outbreak threshold and recovery threshold. The outbreak threshold point is the point of intersection of the susceptible and infected plots in Fig. 6.1. It is defined as the point where the number of infected patients is equal (within a certain threshold) to the number of susceptible patients. The second parameter is the recovery threshold point, which is the point of intersection of the infected and recovered plots in Fig. 6.1. It is defined as the point where the number of recovered patients is equal (within a certain threshold) to the number of infected patients. Figure 6.3 shows how the recovery threshold point varies as a function of β. In the figure, we see as the value of β and I is made 50, the recovery threshold point rapidly decreases. Smaller values of recovery threshold value indicate that the disease recovery starts early. Mathematically, also this makes sense, because $\beta = \frac{k_2}{k_1}$ and larger values of β indicate that the recovery rate is greater than the infection rate.

The recovery threshold point curve can be analyzed by curve fitting techniques. Four different types of curve fitting techniques are evaluated, to examine which one is the best fit. These four types are exponential fit, power fit, cubic spline fit, and B-spline fit. From Fig. 6.4, power fit and the B-spline fit techniques are most accurately representing the recovery threshold point graph. The exponential fit technique does not represent the recovery threshold graph accurately and results in a residual error of 26. Power fit data is generally represented by the equation $y = x^{-\tau}$ with τ value equal to 0.4. Residual error in case of power fit curve is around 5. Between the cubic spline and B-spline techniques, the B-spline technique better represents the graph, with a residual error of 0.34. Hence, the recovery threshold data is best represented by a power fit curve or a B-spline curve. Figure 6.5 shows the recovery threshold graph, only with power fit and B-spline curve fit.

Nodes are assumed to spread disease to each other, only if they are in direct contact with each other. The near-neighbor network simulation is governed by

$$n(x, y) = n(x+1, y) \cup n(x-1, y) \cup n(x, y+1) \cup n(x, y-1) \qquad (6.4)$$

where n is the node, and $n(x, y)$ denotes the coordinate of the node n.

In real-world scenarios, it is often possible that an infected node establishes connections with other nodes in the network, which are not in direct contact. Once this link is established, the infected node is capable to spread the disease to the

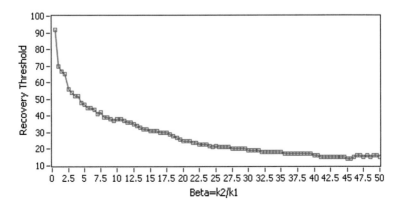

Fig. 6.3 Recovery threshold point versus beta

Fig. 6.4 Recovery threshold point with curve fitting

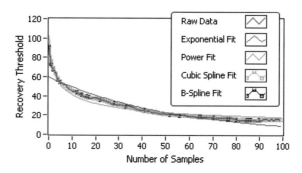

Fig. 6.5 Recovery threshold point with power fit and B-spline fit

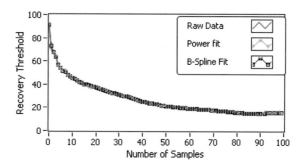

connected nodes. Once the disease spreads to the connected node, it will in turn spread it to the nodes connected to it. This is known as spatial behavior of the nodes. Figure 6.6 shows how the information lifetime varies as a function of the number of connections, an infected node has. As the number of connections increases, we see that the information lifetime increases. In other words, the disease will stay in the network longer (Pan and Sinha 2008). The node with more number of connections is thus capable of controlling the spread of disease. A very important aspect governing the spread of communicable disease is illustrated by the dynamics cited in Gade and Sinha (2005), where the authors compare spread of Ebola virus against HIV virus. People infected with HIV virus live for a long time, and hence have more opportunities to come in contact with other people and consequently

Fig. 6.6 Information lifetime as a function of k_1 and k_2

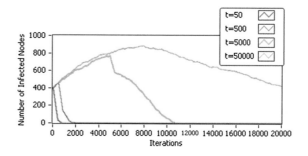

spread the disease. On the contrary, Ebola virus is deadlier than the HIV virus, but the number of people killed by Ebola virus is much smaller as compared to those killed by HIV. This is because people infected by Ebola virus die very quickly, thus not giving the virus an opportunity to spread.

Communicable disease models can be studied for information propagation. Besides this, we can adapt this feature in learning the epidemiology of computer viruses, in routing, in complex networks, etc. (Kephart and White 1991; Vahdat and Becker 2000; Chen and Chen 2010). In the next section, one of the interesting problem rumors spreading in a social network is discussed having both communicable disease model and immune system as the major ingredients (Tsuchiya and Kukino 2004).

6.2   A System for Building Immunity in Social Networks

The rapid evolution of information systems and mobile technology has enabled millions of people to share and express themselves on a common platform, i.e., online social network (Choras et al. 2013). However, this rapid proliferation has aroused privacy concerns as well (Song et al. 2012). Millions of people share information without even knowing where the information reaches eventually (Dinh et al. 2012). One of the challenging problems in this domain is to quickly and effectively stop the spread of malicious information in a social network.

Social networks are susceptible to the rapid spread of malicious information, commonly referred to as rumors (Doerr et al. 2012; Barabasi 2009). Such malicious information often spreads rapidly through the network and can be harmful, if not stopped quickly.

It can be derived that one can draw inspiration from two well-established topics in the area of bio-inspired computing: one is the spread of communicable diseases in human population and second is how human body builds immunity against diseases. Figure 6.7 shows the what, who, and how analogy between the biological systems and the social networks, when it comes to controlling the spread of malicious information in social networks.

In case of social networks, information spreads in two ways. The first way is through near-neighbor effects. A two-dimensional grid is assumed, as shown in Fig. 6.8. Let us consider node A as the node where the disease first originates. Disease will first spread locally, within a sphere of influence denoted by B governed by Eq. (6.4). The number of infected nodes G in this region after a particular amount of simulation time T is counted. Second way in which information spreads depends on the number of connections K the originating node has, with other nodes in the network. In this particular example, let us assume that node A has seven connections. So it is capable of spreading this information to seven other nodes in the system. Each of these seven nodes then creates a local sphere of influence, as depicted by the letters C, D, E, F, G, H, and I. In each of these spheres of influence, information spreads using the near-neighbor effect.

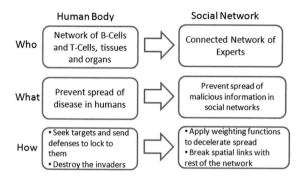

Fig. 6.7 Mapping social network with biological systems

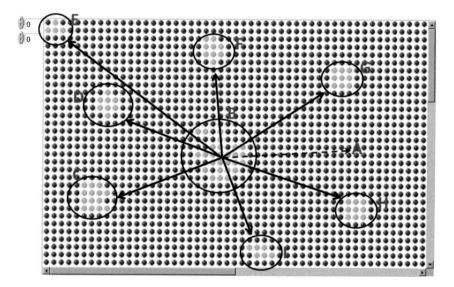

Fig. 6.8 Information spread of a node with multiple connections

A machine learning mechanism can be established where the social network first identifies a set of expert nodes in the system given by Rathore and Samant (2012). These expert nodes are identified based on the study of how they impact the information lifetime as described in the previous section. Initially, a random number of connections for each node are chosen. For each node, the number of infected nodes after a certain time duration T is counted. A threshold value θ_G is chosen. If the number of infected nodes G_i for a particular node i is greater than θ_G, node i is flagged as an expert node. The pseudocode for the algorithm is given in Algorithm 6.1.

Next, a system has a connected system of expert nodes that can effectively sign a weighting function, which controls the spread of malicious information. Doing so builds immunity in the system, which protects the social network from breakdown. Researchers have given various approaches to find such types of experts nodes or influential nodes in social media (Ilyas and Radha 2011; Kempe et al. 2005). The pattern of social media in terms of flow of information is presented in Leskovec (2011).

Algorithm 6.1: Identify Highly Connected Nodes

Input: Number of nodes N
Output: Highly connected nodes
1   for i=0 to N do
2   K = 10*rand();
3   G =countNumInfectedNodes(i, K, T);
4   if $\theta_G \geq G$ then
5   count= count+1;
6   highlyConnectedNodes[count] = i;
7   Return count and highly connected nodes.

6.3   Bio-inspired Solutions for Social Networks

One of the greatest and revolutionary technologies in today's era is social networking. People around the globe stay connected with each other and exchange their well-being online. By the present online networking, not only professionally the researchers collaborate with each other, but also socially the people connect with each other. Bio-inspired solutions have also made a mark in online social network in solving issues like searching, clustering, data diffusion, etc.

The work presented by Rivero et al. (2011) presents novel idea of searching people online through ant colony optimization technique. The online network is a system of connections where people can be represented as nodes of the graph and the relationship between them can be represented as edges. Searching people online requires passing through chain of connections to reach to the desired person. The overall structure of online social network is balanced and personalized. Any relationship can be conceived either as something positive or as something negative. Consider a social network of a set of three people, i.e., A, B, and C. There can be the following scenarios as shown in Fig. 6.9.

(1)   Suppose B and C both are friends with A. It is natural that B and C will also mutually trust each other. This type of relationship is balanced.
(2)   Now, consider both B and C are not friends with A. In a similar manner, it is likely that both B and C can trust each other.

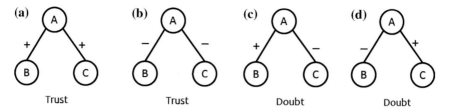

Fig. 6.9 Structural balance

(3) The problem appears when one is a friend and other is not. For instance, B is friend with A and C is not a friend of A. In this case, B and C cannot be friends with each other. This type of relationship is unbalanced.

(4) Likewise, C is a friend of A and B is not a friend of A. In this case too B and C cannot be friends with each other. Suspicion and doubt occurs in this case.

Additionally, online connections are transitive. Let us say, A and B both know each other very well and are best friends. Now, consider B has a friend named C whom A has never met. But, since A knows B so well and trusts B's choices in making friends, A may trust C to a certain extent even though they might have never met. This is also called as *connection propagation* (Guha et al. 2004). See Fig. 6.10.

Now let us say, C has a friend named D whom neither A nor B knows well. Here A is indirectly connected to D. Accordingly, as the link among nodes grows longer, chaining of relationships occurs.

The work presented by Rivero et al. (2011) searches paths in the graphs using concepts such as food (characteristic associated with graph of high centrality) and food odor (used to create areas of odor around the nodes with food to make it easier).

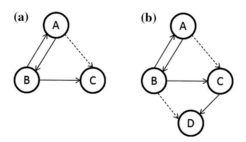

Fig. 6.10 Transitive connections: **a** A is friends with B, B is friends with C, implying A can be friends with C; **b** Chaining of relationships

Issues such as data diffusion and clustering (one of the major social networking problems) can be joined with ethological ways to deal with social conduct in creatures, for example, firefly synchronization (Biojic et al. 2012). Furthermore, identifying influential nodes is a key issue in online networks. Betweenness centrality (Bader et al. 2007) and closeness centrality (Okamoto et al. 2008) do not consider the location of nodes in the networks, and semi-local centrality (Chen et al. 2012), LeaderRank (Li et al. 2014), and PageRank (Mislove et al. 2007) approaches can be just applied in unweighted networks. The work is exhibited by Gao et al. (2013) where a bio-inspired centrality measure model is proposed, which consolidates the Physarum centrality with the K-shell index obtained by K-shell decomposition analysis (Carmi et al. 2007), to identify influential nodes in weighted networks. Then, the susceptible infected (SI) is utilized to assess the execution.

An ad hoc social network (ASNETs) is another branch which investigates social properties of hubs/nodes in correspondences. The use of different social applications in an asset rare environment and the dynamic way of the system make superfluous congestion that may debase the nature of administration significantly. Conventional methodologies use drop tail or random early detection (Athuraliya 2001) techniques to drop data packets from the intermediate node queue. Nonetheless, as a result of the social's inaccessibility properties, these techniques are not suitable for ASNETs. The work exhibited by Liaqat et al. (2014) proposes a bio-inspired packet dropping (BPD) algorithm for ASNETS. BPD copies the coordinating technique of receptors and epitopes in resistant frameworks to recognize congestion. The drop likelihood settings rely on upon the determination of information bundles, which depends on node fame level. BPD chooses the most organized hub through social properties, which is propelled by the B-cell incitement in immune system frameworks. To reasonably organize information bundles, two social properties are utilized: (1) similarity and (2) closeness centrality between hubs.

6.4   Summary

Nature provides amazing inspiration to solve many of the challenging problems faced in social networks today. One of these challenging problems is to quickly and effectively stop the spread of malicious information in a social network. In this chapter, a system for building immunity in social networks using a community of highly connected nodes is presented. This chapter derives inspiration from two naturally occurring phenomena namely epidemic spreading and immune systems. Researchers are engaged in utilizing this fact in a social network which is decentralized, dynamic, and self-governed for more optimum results.

References

Athuraliya, S., Low, S. H., Li, V. H., & Yin, Q. (2001). REM: active queue management. *IEEE Network, 15*(3), 48–53.

Bader, D. A., Kintali, S., Madduri, K., & Mihail, M. (2007). Approximating betweenness centrality. In *Algorithms and models for the web-graph* (pp. 124–137). Berlin: Springer.

Barabási, A. L. (2009). Scale-free networks: A decade and beyond. *Science, 325*(5939), 412.

Bojic, I., Lipic, T., & Podobnik, V. (2012). Bio-inspired clustering and data diffusion in machine social networks. In *Computational social networks* (pp. 51–79). London: Springer.

Carmi, S., Havlin, S., Kirkpatrick, S., Shavitt, Y., & Shir, E. (2007). A model of Internet topology using k-shell decomposition. *Proceedings of the National Academy of Sciences, 104*(27), 11150–11154.

Chen, P. Y., & Chen, K. C. (2010). Information epidemics in complex networks with opportunistic links and dynamic topology In *Global Telecommunications Conference (GLOBECOM 2010)* (pp. 1–6).

Chen, D., Lü, L., Shang, M. S., Zhang, Y. C., & Zhou, T. (2012). Identifying influential nodes in complex networks. *Physica A: Statistical Mechanics and its Applications, 391*(4), 1777–1787.

Choraś, M., Manso, M., Puchalski, D., Kozik, R., & Samp, K. (2013). Online social networks: Emerging security and safety applications. In *Image Processing and Communications Challenges 4* (pp. 291–302). Berlin: Springer.

Dinh, T. N., Shen, Y., & Thai, M. T. (2012). The walls have ears: Optimize sharing for visibility and privacy in online social networks. In *Proceedings of the 21st ACM International Conference on Information and Knowledge Management* (pp. 1452–1461).

Doerr, B., Fouz, M., & Friedrich, T. (2012). Why rumors spread fast in social networks. In *Magazine Communications of the ACM*.

Gade, P. M., & Sinha, S. (2005). Dynamic transitions in small world networks: Approach to equilibrium limit. *Physical Review E-Statistical, Nonlinear and Soft Matter Physics 72*(5), 052903_1–052903_4.

Gao, C., Lan, X., Zhang, X., & Deng, Y. (2013). A bio-inspired methodology of identifying influential nodes in complex networks.

Guha, R., Kumar, R., Raghavan, P., & Tomkins, A. (2004). Propagation of trust and distrust. In *Proceeding of International Conference on World Wide Web* (Vol. 13, pp. 403–412).

Goffman, W., & Newill, V. A. (1964). Generalization of epidemic theory. *Nature, 204*(4955), 225–228.

Hethcote, H. W. (1976). Qualitative analyses of communicable disease models. *Mathematical Biosciences, 28*(3), 335–356.

Ilyas, M. U., & Radha, H. (2011). Identifying influential nodes in online social networks using principal component centrality. In *2011 IEEE International Conference on Communications (ICC)* (pp. 1–5).

Jesan, T., Menon, G. I., & Sinha, S. (2010). Epidemiological dynamics of the 2009 influenza A (H1N1) v outbreak in India. *arXiv preprint* arXiv:1006.0685.

Kempe, D., Kleinberg, J., & Tardos, É. (2005). Influential nodes in a diffusion model for social networks. In *Automata, languages and programming* (pp. 1127–1138). Berlin: Springer.

Kephart, J. O., & White, S. R. (1991). Directed-graph epidemiological models of computer viruses. In *1991 IEEE Computer Society Symposium on Research in Security and Privacy, Proceedings* (pp. 343–359).

Leskovec, J. (2011). Social media analytics: tracking, modeling and predicting the flow of information through networks. In *Proceedings of the 20th International Conference Companion on World Wide Web* (pp. 277–278). ACM.

Li, Q., Zhou, T., Lü, L., & Chen, D. (2014). Identifying influential spreaders by weighted LeaderRank. *Physica A: Statistical Mechanics and its Applications, 404*, 47–55.

Liaqat, H. B., Xia, F., Yang, Q., Xu, Z., Ahmed, A. M., & Rahim, A. (2014). Bio-inspired packet dropping for adhoc social networks. *International Journal of Communication Systems*.

Mislove, A., Marcon, M., Gummadi, K. P., Druschel, P., & Bhattacharjee, B. (2007). Measurement and analysis of online social networks. In *Proceedings of the 7th ACM SIGCOMM Conference on Internet Measurement* (pp. 29–42).

Okamoto, K., Chen, W., & Li, X. Y. (2008). Ranking of closeness centrality for large-scale social networks. In *Frontiers in algorithmics* (pp. 186–195). Berlin: Springer.

Pan, R. K., & Sinha, S. (2008). Modular networks with hierarchical organization: The dynamical implications of complex structure. *Pramana, 71*(2), 331–340.

Rathore, H., & Samant, A. (2012). A system for building immunity in social networks. In *2012 Fourth World Congress on Nature and Biologically Inspired Computing (NaBIC)* (pp. 20–24).

Rathore, H., Ranwa, S., & Samant, A. (2012), Modular network effects on communicable disease models. In *2012 Sixth Asia Modelling Symposium (AMS)* (pp. 126–131).

Rivero, J., Cuadra, D., Calle, F. J., & Isasi, P. (2011). A bio-inspired algorithm for searching relationships in social networks. In *2011 International Conference on Computational Aspects of Social Networks (CASoN)* (pp. 60–65). IEEE.

Song, Y., Karras, P., Nobari, S., Cheliotis, G., Xue, M., & Bressan, S. (2012). Discretionary social network data revelation with a user-centric utility guarantee. In *Proceedings of the 21st ACM International Conference on Information and Knowledge Management* (pp. 1572–1576).

Tsuchiya, T., & Kikuno, T. (2004). An adaptive mechanism for epidemic communication. In *Biologically inspired approaches to advanced information technology.* (pp. 306–316). Berlin: Springer.

Vahdat, A., & Becker, D. (2000). *Epidemic routing for partially connected ad hoc networks* (p. 18). Technical Report CS-200006, Duke University.

Chapter 7
Artificial Neural Network

Abstract Study of artificial neural network (ANN) is a branch of machine learning and data mining. They are a group of measurable learning models inspired by biological neural networks, i.e., brain. The system is utilized to gauge or estimate capacities that can rely upon a substantial number of inputs which are obscure. ANNs are for the most part introduced as frameworks of interconnected "neurons" which trade messages between one another. The associations have numeric weights that can be tuned in view of experience, making neural networks versatile to inputs and fit for learning. The chapter provides details on the ANN and how these frameworks have tackled numerous issues for computer engineers.

7.1 Introduction

An Artificial Neural Network (ANN) is a system of numerous simple processors called "neurons". These processors are connected by communication channels which as a rule convey numeric information, encoded by different means (Pérez-Uribe and Sanchez 2008). The neurons generally operate only on their local data and on the inputs they receive via the connections as shown in Fig. 7.1.

A class of statistical models might usually be called "neural" if it has the accompanying qualities (Wikipedia 2015):

1. It contains sets of adaptive weights, i.e., numerical parameters that are tuned by a learning calculation
2. Has the capability of approximating nonlinear functions of their inputs.

The adaptive weights can be considered as association qualities between neurons, which are actuated amid preparing and expectation. The system's learning is accomplished by giving samples to the system and adjusting to the nearby information put away by the neurons while attempting to lessen the error capacity. Some different routines depend on self-association or experimentation learning.

Some ANN are models of biological neural networks and some are not, but rather truly, a significant part of the motivation for the field of ANNs originated from the yearning to create artificial systems capable for canny calculations like those that

© Springer International Publishing Switzerland 2016
H. Rathore, *Mapping Biological Systems to Network Systems*,
DOI 10.1007/978-3-319-29782-8_7

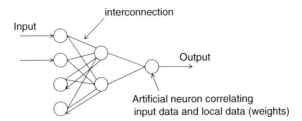

Fig. 7.1 An artificial neural network

the human mind routinely performs. Like other machine learning strategies—
frameworks that gain from information—neural systems have been utilized to tackle
a wide mixed bag of assignments that are difficult to illuminate utilizing conventional rule-based programming, including computer vision and speech recognition.

Enlivened by biological neural networks, ANN is greatly parallel figuring
frameworks comprising of a great degree huge number of basic keen processors with
numerous interconnections. ANN models endeavor to utilize some authoritative
standards believed to be utilized as a part of the human cerebrum. The evolution of
human cerebrum has given numerous opportunities such as (Jain et al. 1996):

- Massive parallelism
- Distributed representation and calculation
- Learning capacity
- Generalization capacity
- Adaptivity
- Inherent logical data handling
- Fault resistance tolerance
- Low energy consumption.

These properties when utilized in networking domain can provide impetus to
solve various research issues.

7.2 Biological Neural Network

In neuroscience, a biological neural system (neural pathway) is a progression of
interconnected neurons whose enactment characterizes a direct pathway. The
interface through which neurons collaborate with their neighbors more often
comprises of a few axon terminals associated by means of synapse (Fig. 7.3) to
dendrites on different neurons. A neuron is arranged in a tree-like structure having
axons and dendrites as shown in Fig. 7.2. The cell body has a nucleus that contains
information about hereditary traits and plasma that holds the molecular equipment
for producing material needed by the neuron.

A neuron receives signals (impulses) from other neurons through its
dendrites (receivers) and transmits signals generated by its cell body along the

Fig. 7.2 Typical structure of neuron

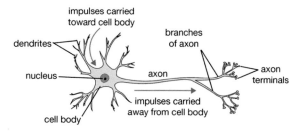

Fig. 7.3 Synapse between neurons

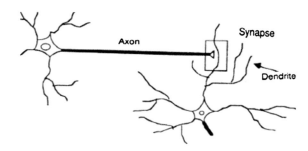

axon (transmitter). At the terminals of axons is the synapse. A synapse is an elementary structure and functional unit between two neurons (one end as axon and the other is the dendrite of other neuron). On the off chance that the information into one neuron surpasses a sure limit, the neuron sends an action potential (AP) at the axon hillock and transmits this electrical sign along the axon. The dependence on history is responsible for learning and memory usage in human brain.

The overall network in human brain comprises of 2^{11} neurons about 2–3 mm thick with a surface area of 2200 cm^2. Each neuron is connected to 10^3–10^4 with other neurons. Human brain is one of the most remarkable unit in human bodies which not only handles fast information processing but also carries out complex perceptual decisions in milliseconds.

Interestingly, a neural circuit is a utilitarian substance of interconnected neurons that has the capacity to manage its own particular movement utilizing a feedback loop (like a control loop in artificial intelligence). Biological neural systems have motivated the configuration of ANNs.

7.3   Artificial Neural Network

The initial model of ANN was proposed by McCulloch and Pitts (1943) which was a binary threshold model (see Fig. 7.4). The model calculated the weighted sum of 'n' input signals x_i, $i = 1, 2 \ldots n$ and generates a binary output if the sum is above a particular threshold 'u' as depicted in the following equation:

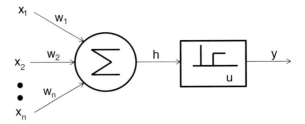

Fig. 7.4 Neuron model by McCulloch–Pitts (*Source* Jain and Mohiuddin 1996)

$$y = \theta\left(\sum_{j=1}^{n} w_i x_i - u\right) \tag{7.1}$$

where

θ is the unit step function at 0
w_i is the synapse weight associated with the respective ith input
u is another weight to check the difference with respect to w.

   The difference between the two leads to positive or negative weights where the positive weight correspond to excitatory synapse and negative weight corresponds to inhibitory synapse.

   ANNs can be depicted as weighted directed graphs where the nodes denote the artificial neuron and edges denote the connection between neuron output and input. It can be classified into two parts:

- Feed forward networks: Graphs without loop
- Recurrent (feedback) networks: Graphs with loop to provide feedback.

   The two models differ in the presence/absence of the feedback. In case of feed forward network, there is a single output for a sequence of inputs. They are a type of static network with a limited memory since they are independent of the previous output. However, recurrent networks are dynamic systems which reiterate the network to produce an optimum output. They require learning memory which updates the results with respect to connection weights to perform task. The performance of the system is dependent on updating weights which is made iteratively for maximal output.

   The learning process is dependent on the environment where the network operates and the weights which are updated with respect to the learning rules. The learning paradigms are of various types (Nelson and Illingworth 1991):

- *Supervised Learning*: It is a type of learning where the mapping is performed between the output and input. It is more of finding a function f for inferring output for given input values. It is generally referred as classification. In machine learning, various classifier techniques are present such as K-means, Gaussian mixture model, naïve Bayes, etc.

- *Reinforcement Learning*: Reinforcement learning is a type of supervised learning that assigns parameters of an ANN, where data is usually not given, but generated by interactions with the environment.
- *Unsupervised Learning*: It is a type of learning which explores the correlations of data rather providing a mapping between the input and output values.
- *Hybrid Learning*: It is the composition of supervised learning and unsupervised learning. Partly weights are determined through supervised learning and partly through unsupervised learning.

7.3.1 Learning Rules in ANN

Learning plays a major role in the ANN and there are fundamental rules for it as explained below.

7.3.1.1 Error Correction Rules

Supervised learning works on the ground of producing output for a given input. However, the output y produced is always not equal to the desired output d. The error $(d - y)$ between the two is used as the error correction rule. A model called perceptron learning model is based on the error correction rule. Backpropagation learning algorithm is based on error correction rule. It consists of a single neuron with adjustable weights and a threshold for a given input (Gallant 1990). Algorithm 7.1 explains the detailed algorithm for perceptron learning.

Algorithm 7.1: Perceptron Learning Rule (Error Correction)

Input: Pattern vector x
Output: Modified connection weights
1   Initialize weights
2   Initialize threshold to small random numbers
3   Present a pattern vector $(x_1, x_2....x_n)^t$ and compute output of neuron
4   Update weights according to following equation
$$w_j(t + 1) = w_j(t) + \alpha(d - y)x_j$$
       Note: d is desired output, t is the iteration number and α $(0.0< \alpha < 1.0)$ is the gain
          (step size)

The input to the neuron is calculated using:

$$v = \sum_{j=1}^{n} w_j x_j - u \qquad (7.2)$$

where

x denotes the pattern vector,
u denotes threshold and
w denotes the weight.

The output is equal to 1 if $v > 0$ or 0 otherwise.
The decision boundary for n dimensional input space is defined by the following linear equation:

$$\sum_{j=1}^{n} w_j x_j - u = 0 \qquad (7.3)$$

Backpropagation algorithm is based on error correction rule as described in Algorithm 7.2 (Buscema 1998). It is an algorithm which trains a given feed forward multilayer neural network for a given set of input patterns with known classifications (Kawaguchi 2000).

Algorithm 7.2: Back-propagation Algorithm

Input: Pattern input vector x
Output: Modified connection weights
1   Initialize weights to small random numbers
2   Repeat until error in the output layer is below a specified threshold or maximum iterations
3   Choose a random input vector
4   Propagate the signal forward through the network
5   Computation of δ_i^L in the output layer (y_i^L) using equation
             $$\delta_i^L = g'(h_i^L)[d_i^u - y_i^L]$$
             Where h_i^L represents the net input to the ith in the layer l^{th} layer and g' is the derivative of the activation function g.
6   Compute deltas for the preceding layers by propagating the error backward which is calculated as below
             $$\delta_i^l = g'(h_i^l)\sum_j w_{ij}^{l+1}\delta_j^{l+1} \quad \text{for } l=L\text{-}1,\ldots\ldots1.$$
7   Update weights according to following equation
             $$\Delta w_{ji}^l = \alpha \delta_j^l y_j^{l-1}$$
             Where α is the learning rate

The supervised error correction learning rule is used in various applications such as function approximation and prediction. It finds its place in performing pattern classification as well (Juang and Katagiri 1992). The unsupervised error correction learning rule is used in data analysis task (Witten and Frank 2005).

7.3.1.2 Boltzmann Learning Rule

Boltzmann learning rule is similar to error correction rule which is used in supervised learning. However, they are slower than error correction learning rule since Boltzmann rule uses Monte Carlo experiments (Arellano and Bond 1991), which are generally slower. The rule uses the state of each neuron and the system output in account for updating the weights. Rather taking the difference between the desired output and the output, difference between the probability distribution is made. Boltzmann learning rule is derived from the thermodynamic principles and information theory (Anderson 1993). The weights are updated with respect to the following equation (Hinton and Salakhutdinov 2006):

$$\Delta w_{ij} = \alpha(\rho_{ij} - \sigma_{ij}) \tag{7.4}$$

where

α is the learning rate
$\rho_{ij},\ \sigma_{ij}$ are the correlations between the states i and j operated in clamped and free running modes, respectively.

Majorly Boltzmann learning rule is used in pattern classification (Duda et al. 2012).

7.3.1.3 Hebbian Rule

The oldest of the learning rule is the Hebbian rule which was given by Hebb in 1949. It states that if the synapse occurs at the both sides of the neuron, the overall synapse increases to twofold (Hebb 2005). The neuron generally observes the orientation selectivity. The weights are updates with respect to the following equation (Martinetz and Schulten 1994):

$$w_{ij}(t+1) = w_{ij}(t) + \alpha y_j(t) x_i(t) \tag{7.5}$$

where

y and x are the output produced by the neuron j and i, respectively,
w is the synapse weight
α is the learning rate.

The supervised Hebbian learning rule is used in data analysis and pattern classification tasks (Shin and Ghosh 1991). The unsupervised Hebbian rule is used in data analysis, data compression (Dony and Haykin 1995).

7.3.1.4 Competitive Learning Rule

Competitive learning rule works on the winner-take-all phenomena, i.e., the best among all the output units, is chosen (DeSieno 1988; Haykin and Network 2004). Using the data correlations clusters are formed which basically groups the input data. Each and every output unit is connected to all the other output units using self-feedback with an excitatory unit. The largest net input $i*$ becomes the winner in this case. The weights of the present learning rule are updated using the following equation:

$$\Delta w_{ij} = \begin{cases} \alpha(x_j^u - w_{i*j}) & i = i^* \\ 0 & \text{otherwise} \end{cases} \tag{7.6}$$

The phenomena of moving the winner unit to the closest of the input is the resultant of the procedure. The learning continuously happens until the learning rate decreases to zero. The learning rule is used in various applications such as image processing, speech processing, data compression, etc. (Fang et al. 1992; Krishnamurthy et al. 1990). Data categorization, data compression, data analysis are majorly the tasks of unsupervised learning (Han et al. 2011).

7.3.2 Types of ANN

An ANN can be described as a parallel system capable of performing those paradigms that linear computing cannot tackle itself. It can be divided into two subcategories of feed forward network and recurrent/feedback networks. The overall classification is as shown in Fig. 7.5.

7.3.2.1 Single Layer and Multilayer Perceptron

Perceptron learning as explained in earlier subsection uses a threshold unit to converge after a finite number of iterations. The learning requires linearly separable ability to produce output hypothesis. An arrangement of one input layer of McCulloch–Pitts neuron feeding forward to one output layer of McCulloch–Pitts neuron is called as a Perceptron model. A single layer feed forward networks consist of one input layer and one output layer of processing unit with no feedback

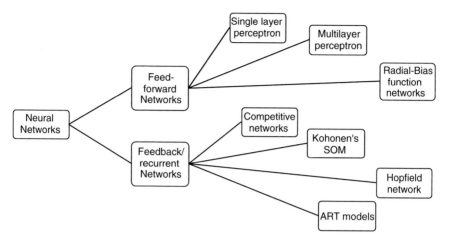

Fig. 7.5 Types of neural network architectures (*Source* Jain et al. 1996)

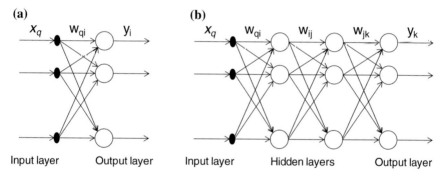

Fig. 7.6 Types of perceptron models **a** Single layer perceptron, **b** Multilayer perceptron

connections. Multilayer feed forward networks contain one input layer along with an output layer having hidden layers additionally as the processing unit as shown in Fig. 7.6.

Multilayer Perceptron model uses threshold function or sigmoid function for the computation. Threshold function produces binary output when the input meets some assigned threshold value as seen in following equation:

$$ y = \begin{cases} 1 & u \geq \theta \\ 0 & u < \theta \end{cases} \tag{7.7} $$

where: θ is the threshold value and u is the activation function.

Sigmoid function introduces nonlinearity in the network and thus can produce any complex decision boundaries and can project Boolean function (Minsky and

Papert 1987). It follows a pattern similar to logistic function and produces a shape S using following function:

$$\text{sig}(t) = \frac{1}{1 + e^{-t}} \tag{7.8}$$

The neural network element produces a linear combination of its input signals and applies a sigmoid function to the result. The backpropagation algorithm as explained in Algorithm 7.2 is a type of multilayer perceptron model which uses gradient descent method to minimize the error cost function as represented by following equation:

$$E = \frac{1}{2} \sum_{i=1}^{p} \left\| y^i - d^i \right\|^2 \tag{7.9}$$

where E is the error cost, y is the output; d is the desired output for a set of p patterns.

7.3.2.2 Radial Basis Function Networks

The Radial Basis Function (RBF) network is a type of multilayer feed forward network having two layers. The output unit implements a weighted sum of hidden unit output. While the input to the RBF network is nonlinear, the output unit is linear in nature. The output depends on the distance of the input from a given stored vector. The layers use Gaussian Kernel as the activation function. However, various activation functions can be used in RBF network. The Gaussian function used in pattern classification can be represented by the following function (Bors 2015):

$$\emptyset_j(x) = \exp\left[-(x - \mu_j)^T \sum_j^{-1} (x - \mu_j) \right] \tag{7.10}$$

$$\text{for } j = 1, \ldots, L.$$

where

x is the input feature vector,
L is the number of hidden units
μ_j and Σ_j are the mean and covariance matrix of the jth Gaussian function.

The mean represents the location and covariance matrix models the shape of the activation function. The output layer implements a weighted sum of hidden unit outputs as represented by the following equation:

$$\omega_k(x) = \sum_{j=1}^{L} w_{jk} \varphi_j(x)$$

(7.11)

$$\text{for } k = 1, \ldots, M$$

where

w_{jk} are the output weights, each corresponding to the connection between hidden unit and output unit. They show the contribution of a hidden unit to the respective output unit

M represents the number of output units.

Since the RBF network employs the usage of large number of hidden layers, the run time of the system is slower than the run time of the multilayer perceptron model.

7.3.2.3 Competitive Networks

Competitive network is a type of unsupervised learning where there exists a competition among nodes to fight for the right to respond to the input values. The network model works on the principle of winner-take-all phenomena. Furthermore, it employs the usage of Hebbian learning rule. The overall phenomena work in the following manner as depicted by Algorithm 7.3. Self-organizing maps (SOM) are a type of competitive network.

Algorithm 7.3: Competitive Network

Input: Neurons with random synaptic weights
Output: Winner neuron
1   repeat for each neuron
2   calculate activation function
3   allow competition among the output neurons via mutual inhibition
4   apply associative Hebbian learning rule
5   normalize the length of the synaptic weights

7.3.2.4 Kohonen's Self-organizing Maps

Kohonen's SOM, a type of recurrent ANN was initially proposed by Teuvo Kohonen in 1982 (Kohonen 1998). They are named as self-organizing since they are not worked under supervision. Furthermore, mapping of weights is required to conform to the input data. A typical Kohonen self-organizing map is shown in

Fig. 7.7 Self-organizing
maps architecture

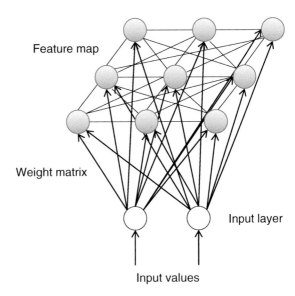

Fig. 7.7. The crux of this network is the feature map, a neuron layer where neurons
organize themselves according to certain input values. The algorithm for the same is
as shown in Algorithm 7.4.

Algorithm 7.4: SOM Learning Algorithm

Input: Pattern input vector x
Output: Modified connection weights
1   Initialize weights to small random numbers and set initial rate and neighborhood
2   Repeat until the output layer is below a specified threshold or maximum iterations
3   Choose a random input vector
4   Select unit (c_i, c_j) with minimum output
         $$\|x - w_{c_i, c_j}\| = min_{ij} \|x - w_{ij}\|$$
5   Update all weights according to following learning rule:
         $$w_{ij}(t+1) = \begin{cases} w_{ij}(t) + \alpha(t)[x(t) - w_{ij}(t)] & if\ (i,j) \in N_{c_i, c_j} \\ w_{ij}(t) & otherwise \end{cases}$$
         Where $N_{c_i, c_j}(t)$ is the neighborhood of the unit (c_i, c_j) at time t
               $\alpha(t)$ is the learning rate
6   Decrease the value of $\alpha(t)$ and shrink the neighborhood $N_{c_i, c_j}(t)$

7.3.2.5 Hopfield Network

Hopfield network is a type of recurrent ANN model, devised by Little and later
popularized by Hopfield (1982). It is a fully connected neural network and the weights

are kept symmetric with no self-connections which states that the weight from neuron i to neuron j is equal to the weight from neuron j to neuron i. It can be seen as a network with associative memory having two phases, viz., storage and retrieval. Each neuron is connected to every other neuron such as all neurons act as input and output as shown in Fig. 7.8. The network comprises of set of neurons and the corresponding set of unit delays to organize the network as a multiple loop feedback system.

The activation values of the Hopfield network are binary $\{+1, -1\}$ or continuously valued {between 0 and 1} in case of continuous network. The activation value is calculated using the following equation (Laferriere 2010):

$$x_i(t+1) = \text{sign}\left(\sum_{j=1}^{n} x_j(t)w_{ij} - \theta_i\right) \qquad (7.12)$$

Where

x is the activation value of n neurons
w is the weight and
θ is the threshold.

The sign function is defined as:

$$\begin{cases} +1 & \text{if } x \geq 0 \\ -1 & \text{otherwise} \end{cases} \qquad (7.13)$$

Furthermore, the weights are updated in two ways, i.e., via synchronously and asynchronously. In asynchronous update, one random neuron is picked and weighted sum of neurons is calculated. In case of synchronous update, the weighted input sums of all neurons are calculated without updating the neurons.

Fig. 7.8 Hopfield network

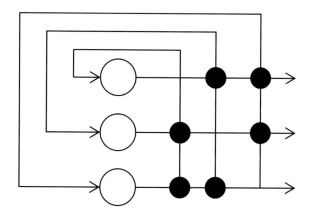

7.3.2.6 Adaptive Resonance Theory Model

The stability-plasticity dilemma is a well-known constraint for artificial and biological neural systems which was observed during competitive learning (Mermillod
et al. 2013). The essential thought is that learning in a parallel and disseminated
framework requires plasticity for the reconciliation of new information, additionally
stability with a specific end goal to keep overlooking the past information. A lot of
plasticity will bring about already encoded information being never remembered,
though an excessive amount of stability will obstruct the effective coding of this
information at the level of the neural connections. Adaptive Resonance Theory
(ART) tries to resolve stability and plasticity dilemma in the following way:

- Stability is attained because previous learning is preserved since the existing
  clusters are not altered
- Plasticity is attained as the new example is incorporated by creating a new
  cluster.

ART is a type of unsupervised learning which was developed by Carpenter and
Grossberg (2010). The model is similar to clustering issue where the distance from
the nearest cluster is computed and updated. The difference lies in the updating
technique where the cluster is assigned only when the distance is very less,
otherwise a new cluster is formed to handle. The procedure to check if the cluster
assignment is efficient a threshold of similarity is introduced between patterns and
clusters. The model is similar to leader follower algorithm (Tamu 2015).

There are various types of ART, namely ART 1 for binary features, ART 2 for
analog features, and ARTMAP which is a supervised version of ART. The algorithm for ART 1 is explained in Algorithm 7.5.

Algorithm 7.5: Adaptive Resonance Theory 1 Learning Algorithm

---

Input: Initialize $w_{ij}=1$ for i, j. Enable all output units.
Output: Modified connection weights
1   Present pattern x
2   Discover the winner unit i* among the enabled output units
$$\overline{w_{i^*}}.x \geq \overline{w_i} \quad \forall i$$
3   Perform vigilance test
$$r = \frac{w_{i^*}\overline{x}}{\sum_i x_i}$$
4   If $r \geq \rho(resonance)$
              Go to step 6
5   Else
              Disable unit i* and go to step 2 until all output units are disabled
6   Update wining weight vector w_{ji^*} enable output units and go to step 1
$$\Delta w_{ji^*} = \theta(V_j - w_{ji^*})$$
7   If all output units are disabled, select one of committed output and set weight as x
8   If the committed output is not present, reject the input pattern.

7.4   Applications of Artificial Neural Networks

ANNs perform complex, dynamic, and intelligent behavior which is being utilized in various applications such as forecasting, data mining, and operating systems processing of task scheduling and resource allocation tasks, etc. The section discusses some of the applications of ANNs.

7.4.1   Pattern Classification

Classification is a crucial task in many machine learning algorithms. Finding a pattern is to assign an input pattern to many prescribed classes. ANN is used in variety of applications such as character recognition (Rogova 1994), speech recognition (Waibel et al. 1989). Besides that, various biological applications are inspired from ANN in resolving blood cell classification (Theera-Umpon and Dhompongsa 2007), EEG waveform classification (Vuckovic et al. 2002). The work done by Looney (1997) provides interesting theory and algorithms that can be utilized by scientists and engineers for solving pattern recognition problems. Furthermore, many useful work linking both pattern recognition and ANN can be studied from (Khan et al. 2001; Specht 1990; García-Pedrajas et al. 2005), Sietsma and Dow 1991).

7.4.2   Clustering

Clustering is a type of unsupervised learning model where unlabelled data set is given labels. Clustering model studied similar patterns and draws a demarcation between the classes so formed. Various clustering applications are used in data mining, data compression, and data analysis. Linking the clustering phenomena and ANN can be utilized in solving many data-related queries. Work done by Craven and Shavlik (1997), Berry and Linoff (1997) provides details on diverse data mining issues and applications. A VLSI neural processor for image data compression using self-organizing networks is studied in Fang et al. (1992). Exploratory multivariate data analysis is studied in Mao and Jain (1995). Besides that exploratory data analysis can be performed using SOM (Kaski 1997). Lowe and Tipping (1996) studied the feed forward neural network and topographical mapping for exploratory data analysis.

7.4.3   Optimization

Optimization is defined as utilization of minimum resources with production of maximal output. Optimization techniques can be utilized in planning, allocation, decision-making, scheduling processes. Traveling salesman problem is a classic example of optimization technique which is linked with Kohonen type of neural network (Fritzke 1994; Favata and Walker 1991; Rosenkrantz et al. 1977).

7.4.4 Prediction and Forecasting

Prediction and forecasting is used for determining the future trends based on present and past data and is mainly used in stock market analysis and weather forecasting. ANNs are used in forecasting applications. The state of art is given in Zhang et al. (1998). A review of literature comparing ANNs and statistical models, particularly in regression-based forecasting, time series forecasting, and decision-making is presented in Hill et al. (1994).

7.5 Summary

ANN can be defined as computational algorithm which is inspired by the structure, process, and learning ability of a human brain. Human brain has interesting characteristics of distributed representation of knowledge, fault tolerance, massive parallelism, generalization, and learning ability. Learning can be done in various manner, viz., supervised, unsupervised, reinforcement and hybrid. The chapter deals with ANNs and the modeling of biological brain with computer network domain.

References

Anderson, J. A. (1993). *Neurocomputing* (Vol. 2). Cambridge: MIT Press.
Arellano, M., & Bond, S. (1991). Some tests of specification for panel data: Monte Carlo evidence and an application to employment equations. *The Review of Economic Studies, 58*(2), 277–297.
Berry, M. J., & Linoff, G. (1997). *Data mining techniques: For marketing, sales, and customer support*. New York: Wiley.
Bors, A. (2015) Introduction to Radial Basis Function (RBF) Networks. https://www-users.cs.york.ac.uk/adrian/Papers/Others/OSEE01.pdf.
Buscema, M. (1998). Back propagation neural networks. *Substance Use and Misuse, 33*(2), 233–270.
Carpenter, G. A. & Grossberg, S. (2010). *Adaptive resonance theory* (pp. 22–35). Springer.
Craven, M. W., & Shavlik, J. W. (1997). Using neural networks for data mining. *Future generation computer systems, 13*(2), 211–229.
DeSieno, D. (1988). Adding a conscience to competitive learning. In *IEEE International Conference on Neural Networks*, (pp. 117–124).
Dony, R. D., & Haykin, S. (1995). Neural network approaches to image compression. *Proceedings of the IEEE, 83*(2), 288–303.

Duda, R. O., Hart, P. E., & Stork, D. G. (2012). *Pattern classification*. New York: Wiley.

Fang, W. C., Sheu, B. J., Chen, O. T., & Choi, J. (1992). A VLSI neural processor for image data compression using self-organization networks. *IEEE Transactions on Neural Networks, 3*(3), 506–518.

Favata, F., & Walker, R. (1991). A study of the application of Kohonen-type neural networks to the travelling salesman problem. *Biological Cybernetics, 64*(6), 463–468.

Fritzke, B. (1994). Growing cell structures—a self-organizing network for unsupervised and supervised learning. *Neural Networks, 7*(9), 1441–1460.

Gallant, S. (1990). Perceptron-based learning algorithms. *IEEE Transactions on Neural Networks, 1*(2), 179–191.

García-Pedrajas, N., Hervás-Martínez, C., & Ortiz-Boyer, D. (2005). Cooperative coevolution of artificial neural network ensembles for pattern classification. *IEEE Transactions on Evolutionary Computation, 9*(3), 271–302.

Han, J., Kamber, M., & Pei, J. (2011). *Data mining: concepts and techniques: Concepts and techniques*. New York: Elsevier.

Haykin, S. & Network, N. (2004). A comprehensive foundation. *Neural Networks*, No. 2.

Hebb, D. O. (2005). *The organization of behavior: A neuropsychological theory*. Psychology Press.

Hill, T., Marquez, L., O'Connor, M., & Remus, W. (1994). Artificial neural network models for forecasting and decision making. *International Journal of Forecasting, 10*(1), 5–15.

Hinton, G. E., & Salakhutdinov, R. R. (2006). Reducing the dimensionality of data with neural networks. *Science, 313*(5786), 504–507.

Hopfield, J. J. (1982). Neural networks and physical systems with emergent collective computational abilities. *Proceedings of the National Academy of Sciences, 79*(8), 2554–2558.

Jain, A. K., Mao, J. & Mohiuddin, K. M. (1996). Artificial neural networks: A tutorial. *Computer* (3), 31–44.

Juang, B. H., & Katagiri, S. (1992). Discriminative learning for minimum error classification [pattern recognition]. *IEEE Transactions on Signal Processing, 40*(12), 3043–3054.

Kaski, S. (1997). Data exploration using self-organizing maps. *Acta Polytechnica Scandinavica: Mathematics, Computing and Management in Engineering Series* 82.

Kawaguchi, K., (2000). Back-propagation learning algorithm. http://www.ece.utep.edu/research/webfuzzy/docs/kk-thesis/kk-thesis-html/node22.html. Accessed on October 19, 2015.

Khan, J., Wei, J. S., Ringner, M., Saal, L. H., Ladanyi, M., Westermann, F., & Meltzer, P. S. (2001). Classification and diagnostic prediction of cancers using gene expression profiling and artificial neural networks. *Nature Medicine, 7*(6), 673–679.

Kohonen, T. (1998). The self-organizing map. *Neurocomputing, 21*(1), 1–6.

Krishnamurthy, A. K., Ahalt, S. C., Melton, D. E., & Chen, P. (1990). Neural networks for vector quantization of speech and images. *IEEE Journal on Selected Areas in Communications, 8*(8), 1449–1457.

Laferriere, A. L. (2010). Hopfield network. http://perso.ens-lyon.fr/eric.thierry/Graphes2010/alice-julien-laferriere.pdf.

Looney, C. G. (1997). *Pattern recognition using neural networks: theory and algorithms for engineers and scientists*. Oxford: Oxford University Press Inc.

Lowe, D., & Tipping, M. (1996). Feed-forward neural networks and topographic mappings for exploratory data analysis. *Neural Computing and Applications, 4*(2), 83–95.

Mao, J., & Jain, A. K. (1995). Artificial neural networks for feature extraction and multivariate data projection. *IEEE Transactions on Neural Networks, 6*(2), 296–317.

Martinetz, T., & Schulten, K. (1994). Topology representing networks. *Neural Networks, 7*(3), 507–522.

McCulloch, W. S., & Pitts, W. (1943). A logical calculus of the ideas immanent in nervous activity. *The Bulletin of Mathematical Biophysics, 5*(4), 115–133.

Mermillod, M., Bugaiska, A. & Bonin, P. (2013). The stability-plasticity dilemma: investigating the continuum from catastrophic forgetting to age-limited learning effects. *Frontiers in Psychology, 4.*

Minsky, M. L., & Papert, S. A. (1987). *Perceptrons—expanded edition: an introduction to computational geometry*. Boston, MA: MIT Press.

Nelson, M. M., & Illingworth, W. T. (1991). *A practical guide to neural nets* (Vol. 1). Reading, MA: Addison-Wesley.

Pérez-Uribe, A. & Sanchez, E. (2008) *Bio-inspired techniques and their application to precision agriculture*. REDS Institute (http://reds.eivd.ch) University of Applied Sciences of Western-Switzerland—EIVD.

Rogova, G. (1994). Combining the results of several neural network classifiers. *Neural Networks, 7*(5), 777–781.

Rosenkrantz, D. J., Stearns, R. E., & Lewis, P. M, I. I. (1977). An analysis of several heuristics for the traveling salesman problem. *SIAM Journal on Computing, 6*(3), 563–581.

Shin, Y. & Ghosh, J. (1991) The pi-sigma network: An efficient higher-order neural network for pattern classification and function approximation. In *IJCNN-91-Seattle International Joint Conference on Neural Networks* (Vol. 1, pp. 13–18).

Sietsma, J., & Dow, R. J. (1991). Creating artificial neural networks that generalize. *Neural Networks, 4*(1), 67–79.

Specht, D. F. (1990). Probabilistic neural networks. *Neural Networks, 3*(1), 109–118.

Tamu, (2015). http://research.cs.tamu.edu/prism/lectures/pr/pr_l16.pdf. Accessed on October 25, 2015.

Theera-Umpon, N., & Dhompongsa, S. (2007). Morphological granulometric features of nucleus in automatic bone marrow white blood cell classification. *IEEE Transactions on Information Technology in Biomedicine, 11*(3), 353–359.

Vuckovic, A., Radivojevic, V., Chen, A. C., & Popovic, D. (2002). Automatic recognition of alertness and drowsiness from EEG by an artificial neural network. *Medical Engineering & Physics, 24*(5), 349–360.

Waibel, A., Hanazawa, T., Hinton, G., Shikano, K., & Lang, K. J. (1989). Phoneme recognition using time-delay neural networks. *IEEE Transactions on Acoustics, Speech and Signal Processing, 37*(3), 328–339.

Wikipedia (2015). Artificial neural network. Accessed on September 23, 2015.

Witten, I. H. & Frank, E. (2005) *Data mining: Practical machine learning tools and techniques*. Morgan Kaufmann.

Zhang, G., Patuwo, B. E., & Hu, M. Y. (1998). Forecasting with artificial neural networks: The state of the art. *International Journal of Forecasting, 14*(1), 35–62.

Chapter 8
Genetic Algorithms

Abstract Genetic algorithms are the heuristic search and enhancement upgrade structures that impersonate the undertaking of natural evolution. It is a machine learning algorithm which comprises a populace of individuals represented by chromosomes (present in individual's DNA). The number of inhabitants in people contends with one another for assets to reach to another era of people, a procedure commonly known as evolution. The people capable of surviving spread their hereditary material. This chapter provides details of how genetic algorithms work and the way they are used in computer networks for solving issues such as sensor networks, traveling salesman problem, etc.

8.1 Introduction

A genetic algorithm (or GA) is a pursuit strategy utilized as a part of registering to discover genuine or approximate solutions for optimization and search problems. These are arranged as worldwide pursuit heuristics. They are a specific class of evolutionary algorithms that utilize strategies raised by evolutionary biology such as *inheritance*, *mutation*, *selection*, and *crossover* (also called recombination). The process starts with a set of population where the fitness of each individual is calculated and based on it; a new modified population is calculated (best fitness) as shown in Fig. 8.1.

Genetic algorithm is a class of computational modeling which explores a large number of possibilities for optimized solutions as shown in Fig. 8.2. It works on the principle of selecting the best and discarding the rest. The genetic algorithm has its foundational roots starting from the work implemented by Holland (1975). In the early 1980s, techniques of genetic algorithm were applied in diverse manners. Genetic programming was another algorithm used to evolve programs to perform tasks developed by Koza (1992).

Things being what they are, there is no thorough meaning of "genetic algorithm" acknowledged by all in the evolutionary computation group that separates GAs from other transformative calculation strategies. In any case, it can be said that most

© Springer International Publishing Switzerland 2016   97
H. Rathore, *Mapping Biological Systems to Network Systems*,
DOI 10.1007/978-3-319-29782-8_8

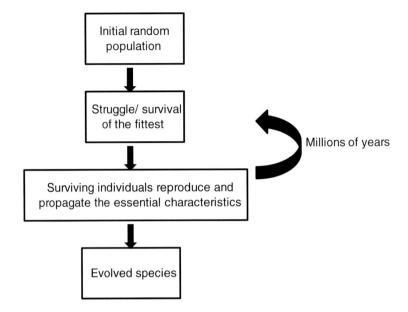

Fig. 8.1 Natural evolution

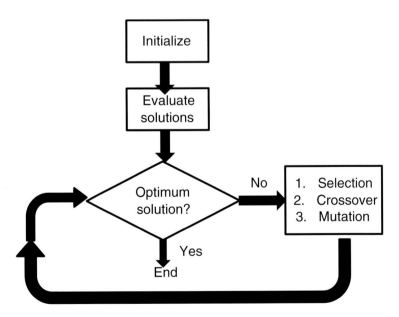

Fig. 8.2 Process of finding optimized solution

routines called "GAs" have at any rate the accompanying components in like manner: populaces of chromosomes, choice as per wellness, hybrid to create new offspring, and transformation of new offspring. Inversion—Holland's fourth component of GAs—is once in a while utilized as a part of today's executions, and its points of interest, if any, are not settled. The chromosomes in a GA populace ordinarily take the type of bit strings. Every locus in the chromosome has two conceivable alleles: 0 and 1. Every chromosome can be considered as a point in the hunt space of applicant arrangements. The GA forms populaces of chromosomes, progressively supplanting one such populace with another. The GA frequently requires a wellness capacity that appoints a score (wellness) to every chromosome in the present populace. The wellness of a chromosome relies on how well that chromosome takes care of the current situation.

8.2 Genetic Algorithms

In the 1950s and 1960s, researchers freely concentrated on evolutionary systems with the thought that evolution could be utilized as an improvement apparatus for designing issues. The thought in each of these frameworks was to advance a populace of applicant solutions for a given issue, utilizing administrators raised by natural genetic variation and natural selection. In the 1960s (Rechenberg 1965) presented "evolution strategies," a system he used to improve real-valued parameters for gadgets, for example, airfoils. Later, Fogel et al. (1966) created "evolutionary programming," a strategy in which candidate solutions, competitor answers for given assignments were spoken of as finite-state machines, which were developed by arbitrarily changing their state—transition charts and selecting the fittest.

Interestingly with evolution strategies and evolutionary programming, Holland's unique objective was not to plan calculations to take care of particular issues, but to formally think about the adaptation as it happens in nature and to create routes in which the components of natural adaptation might be imported into computer system frameworks. Holland's (1975) book "Adjustment in Natural and Artificial Systems" exhibited the hereditary calculation as a reflection of organic advancement and gave a hypothetical structure for adjustment under the GA. Holland's GA is a strategy for moving from one populace of chromosomes (e.g., series of ones and zeros, or "bits") to another populace by utilizing a sort of "natural selection" together with the genetics-inspired administrators of *crossover*, *mutation*, and *inversion*.

Every individual comprises rules which determine its way of being built. Cell is the basic unit of living organisms. Each cell comprises a set of chromosomes. Chromosomes are strings of DNA which serve as a process for the living organism. Each chromosome consists of "genes" or block of DNA. Each gene encodes a specific that speaks of the attribute of an individual, for example, color of eyes. Each gene having an instance of a specific "allele" defines the trait of the individual. The selection picks those chromosomes in the populace that will be permitted to recreate, and overall the fitter chromosomes create more posterity than the less fit ones. Crossover exchanges subparts of two chromosomes by generally copying the natural recombination between two single-chromosome ("haploid") organisms; mutation arbitrarily changes the allele estimations of a few areas in the chromosome, and inversion inverts the segment of the chromosome, subsequently revamping the request in which genes are displayed.

The complete gathering of hereditary material (all chromosomes taken together) is known as the organism's genome. The term genotype alludes to the specific arrangement of qualities contained in a genome. Two people that have indistinguishable genomes are said to have the same genotype. The genotype gives ascend, under fetal and later advancement, to the organism's phenotype—its physical and mental qualities, for example, eye color, tallness, cerebrum size, and knowledge. Living beings whose chromosomes are displayed in sets are called diploid; living beings whose chromosomes are unpaired are called haploid. In nature, most sexually duplicating species are diploid, including people, which have 23 sets of chromosomes in each substantial (non-germ) cell in the body. Amid sexual propagation, recombination (or crossover) happens: in every parent organism, qualities are traded between every pair of chromosomes to frame a gamete (a solitary chromosome), and afterward gametes from the two folks pair up to make a full arrangement of diploid chromosomes. In haploid sexual multiplication, qualities are traded between the two folks' single-strand chromosomes. Posterity is liable to transformation, in which single nucleotides (basic bits of DNA) are changed from parent to offspring, the progressions regularly coming about because of replicating mistakes. The wellness of a living being is regularly characterized as the likelihood that the life form will live to reproduce (feasibility) or as an element of the quantity of posterity the living being has (fertility).

As a whole, the natural evolution usually initiates from a populace of arbitrarily created people and produces eras of generation. In each era of generation, the fitness of every individual in the population is assessed; multiple individuals are chosen from the present population (based on their fitness) and adjusted to shape another populace.

In genetic algorithms, the term chromosome regularly alludes to an applicant answer for an issue, frequently encoded as a bit string. The *genes* are either single bits or short blocks of adjacent bits that encode a specific component of the candidate solution (e.g., in the context of multiparameter function optimization, the bits encoding a specific parameter are considered to be a gene). An allele in a bit string is either 0 or 1; for larger alphabets more are conceivable at every locus. Crossover comprises exchanging genetic material between two single-chromosome haploid parents. Mutation comprises flipping the bit at a haphazardly picked locus (or, for bigger letters in order, supplanting the image at an arbitrarily picked locus with an arbitrarily picked new image). Most applications of genetic algorithms employ haploid individuals, particularly, single-chromosome people. The genotype of a person in a GA utilizing bit strings is basically the arrangement of bits in that individual's chromosome. There is no notion of "phenotype" in the context of GAs, although recently many workers have experimented with GAs in which there is both a genotypic level and a phenotypic level (e.g., the bit-string encoding of a neural network and the neural network itself).

The basic genetic algorithm is given in Algorithm 8.1.

Algorithm 8.1: Basic Genetic Algorithm

Input: A large "population" of randomly generated "attempted solutions" to a problem
Output: New population

1   repeat until satisfied solution is not attained or time out
2   Evaluate each of the attempted solutions
3   (probabilistically) keep a subset of the best solutions based on
                     Selection
                     Crossover
                     Mutation
                     Accepting
4   Use these solutions to generate a new population

8.3   Encoding

Encoding is generally performed before genetic algorithm so that the computer framework is able to process it. A usual way is to represent the information in the form of binary 0's and 1's. There are diverse ways of encoding such as representing the solution in the form of integers or real-valued numbers.

- *Binary encoding* is a way of processing the problem and solution in a sequential set of 0's and 1's. For instance, representation of chromosome can be made in the form of 10101000.
- *Value encoding* is yet another way of encoding strategy where real values are used such as chromosome can be represented in the form of 2.3456, NFFISJIJGIOFJGJDFOHFFD.
- *Permutation encoding* is utilized in ordering problems such as traveling salesman problem and eight queen's problem. Here the chromosome is a set of numbers that represents position in the set such as 153452346.
- *Tree encoding* is the other way of representing the evolving programs and expressions. The representation is made in the form of a tree having a root node with a number of leaf nodes with levels.

8.4   Genetic Algorithm Operators

As explained earlier, the simplest form of genetic algorithm involves three types of operators: selection, crossover (single point), and mutation for the production of new era.

8.4.1   *Selection*

Selection operator selects and picks chromosomes in the population for reproduction based on the fitness function. The fitter the chromosome, the more times it is likely to be selected to reproduce. The fitness level provides more optimum solution expected in concordance with the desired result. Selection strategy chooses the chromosomes to apply survival of the fittest mechanism on them. The selection procedure is of various types, roulette wheel selection (Lipowski and Lipowska 2012), stochastic universal selection (Miller and Goldberg 1996), proportionate selection (Bridges and Goldberg 1987), genitor selection (Whitley 1989), ranking selection, and tournament selection (Chakraborty and Chakraborty 1997). Roulette wheel selection strategy is based on the wheel architecture where the individual has the maximum fitness. Figure 8.3 shows the roulette wheel selection strategy for $n = 8$ individuals. For the present case the 4th individual has maximum fitness value of 30 %. Wheel is spun $n = 8$ times and fitness is calculated using the following equation:

$$\text{Average fitness} = F'F_j/n \qquad (8.1)$$

where

F fitness
n number of individuals

Probability of selecting the ith string is calculated using:

$$p_i = F_i / \left(\sum_{j=1}^{n} F_j \right) \qquad (8.2)$$

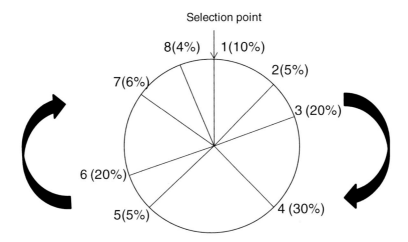

Fig. 8.3 Roulette wheel selection

Expected outcome is calculated using:

$$\text{Count} = (n = 8) \times p_i$$

Comparative analysis of the selection models is given in Goldberg and Deb (1991), Zhong et al. (2005). Linear ranking (Baker 1985) and tournament ranking (Brindle 1981) showed the same performance and maintained strong growth under normal conditions. Tournament selection strategy converges faster than the roulette wheel selection strategy. Proportionate selection is less effective than linear and tournament ranking since it maintains a less steady pressure during convergence. Furthermore, genitor selection shows premature convergence.

8.4.2 Crossover

Crossover operator crosses two parent individuals and produces offspring. If the crossover is not performed then the offspring is exactly the replica of the parent individual. Crossover is performed by randomly choosing a locus and exchanging the subsequences before and after that locus between two chromosomes to create two offspring. The crossover operator roughly mimics biological recombination between two single-chromosome (haploid) organisms. Crossover operators are of various types such as one-point, two-point, uniform, arithmetic, and heuristic crossover.

One-point crossover selects a random point after which it changes the bit sequence. The mark | depicts the demarcation point after which the bits are changed. For instance, the strings 100|00100 and 111|11111 could be crossed over after the third locus in each to produce the two offspring 10011011 and 11100000.

Two crossover points changes the bit sequence between two points. For instance, the strings 100|001|00 and 111|111|11 could be crossed over in each to produce the two offspring 10000100 and 11100011.

Uniform crossover produces the offspring with some mixing ratio of the parent individual, i.e., probability is assigned to the parent individual for its contribution in producing the offspring. For instance, if the mixing ratio is 0.5, half of the contribution is from one parent and the other half from another parent. The strings 1000010010000100 and 1111111111111111 could be crossed over in each to produce the two offspring $1_1 1_2 1_2 0_1 0_1 1_2 1_2 1_2 1_1 0_1 1_2 0_1 1_2 1_1 0_1 1_2$ and $1_2 0_1 0_1 1_2 1_2 1_1 0_1$ $0_1 1_2 1_2 0_1 1_2 0_1 1_2 1_2 0_1$, where the subscript represents the parent.

Arithmetic crossover produces the two offspring according to the weighting function (w) as defined by the following functions:

$$\text{Offspring}_1 = w * \text{Parent}_1 + (1 - w) * \text{Parent}_2 \tag{8.3}$$

$$\text{Offspring}_2 = (1 - w) * \text{Parent}_1 + w * \text{Parent}_2 \tag{8.4}$$

For instance, consider parents represented by 4 gene values as $parent_1 = (0.1)$ $(1.2)(2.3)(0.4)$ and $parent_2 = (0.3)(1.4)(0.3)(0.7)$ with $w = 0.3$. Offspring are calculated using Eqs. (8.3) and (8.4) as $Offspring_1 = (0.24)(1.34)(0.9)(0.61)$ and $Offspring_2 = (0.16)(1.26)(1.7)(0.49)$.

Heuristic crossover uses the fitness values and random number r (between 0 and 1) to determine the offspring according to the following equation:

$$Offspring_1 = Parent_{best} + r * (Parent_{best} - Parent_{worst}) \qquad (8.5)$$

$$Offspring_2 = Parent_{best} \qquad (8.6)$$

8.4.3 Mutation

Mutation operator randomly flips some of the bits in a chromosome to produce a better result in comparison to the result produced by previous operator. For example, the string 00111100 might be mutated in its second position to produce 01111100. Mutation operator is of various types such as flip-bit, boundary, nonuniform, uniform, and Gaussian. Boundary mutation replaces the bit with upper bound or lower bound of the gene. Uniform operator adds a uniform value between the upper and lower bound specified. Nonuniform operator works the other way round, for allowing mutation not to stagnate at early stages of evolution. Gaussian operator adds Gaussian distributed value.

8.5 Applications of Genetic Algorithms

Genetic algorithms applications include game theory, control gas pipeline missile evasion, designing aircraft, keyboard configuration, security, and trajectory planning in robotics.

Genetic algorithm is used in a variety of optimization problems such as traveling salesman problem where the minimum distance is covered with the maximum coverage of cities. Besides, genetic algorithm is also used in job shop scheduling, and video and sound quality optimization. Another basic utilization of GAs is function optimization, where the objective is to locate an arrangement of parameter values that amplify, say, a complex multiparameter function. As a straightforward sample, one might need to maximize the real-valued one-dimensional capacity (Riolo 1992). Here the hopeful arrangements are estimations of y, which can be encoded as bit strings speaking to genuine numbers. The wellness count interprets a given piece string x into a genuine number y and after that assesses the function. The wellness of a string is the function value. As a non-numerical illustration, consider the issue of discovering an arrangement of 50 amino acids that will overlay to a fancied three-dimensional protein structure. A GA could be connected to this issue via looking a populace of hopeful arrangements, each encoded as a 50-letter string, for example (Melanie 1999),

IHCCVASASDMIKPVFTVASYLKNWTKAKGPNFEICISGRTPYWDNFPGI,

where every letter speaks to 1 of 20 conceivable amino acids. One approach to characterize the wellness of a competitor arrangement is as the negative of the potential vitality with respect to the desired structure.

Automatic programming is yet another field where genetic algorithm is used for designing a computational model such as cellular automata and sorting networks. Furthermore, it is also used in diverse machine learning applications such as classification and prediction. It is also used in neural network for learning the classifier function. Immune system aspects such as somatic mutations are also designed taking in the genetic algorithm aspect. Population genetic model, ecological model, and social network models (Pizzuti 2008) also use various genetic algorithm aspects.

Bio-inspired self-adaptive rate control for multi-priority data transmission for wireless LANs was proposed by Yao et al. (2014). Mapping between wireless LAN and ecosystem can be very well represented as shown in Table 8.1.

The work uses Lotka-Volterra Model (Zhu and Yin 2009), which is studied for natural evolution of species with respect to the resources available in the ecosystem. The equation for the same is given as

$$\frac{\mathrm{d}x_i(t)}{\mathrm{d}t} = x_i(t)\left(r_i - r_i \sum_{j=1, j\neq i}^{n} a_{ij}x_i(t)\right) \qquad (8.7)$$

where

$x_i(t)$ represents the population size of species i at time t;
r_i is the growth rate intensity of species i;
n is the number of species in a biological system, and
a_{ij} is interspecies competition coefficient, which is defined as the competition effect from species j to species i.

The scheme guarantees 93 % progress in bandwidth utilization compared to the existing EDCA protocol. Furthermore, it also guarantees quality of service and service differentiation which provides better data transmission.

Table 8.1 Mapping between ecosystem and WLAN

Ecosystem	Wireless LANs
Multiple species	Multiple categories of data flows
Resources	Network bandwidth
Population size	Sending rate
Position of species	Priority allocated to each category
Competition among species	Competition among species

8.6   Summary

Genetic algorithm produces better solutions because of the interesting characteristics it possesses. It uses three major components, viz, selection, crossover, and
mutation which aid in reaching an optimized solution. It finds its application in
many networking issues such as traveling salesman problem, video and sound
quality optimization, immune systems, social networks, etc. This chapter provides
an understanding of biological genetic material and how it is comprehended in the
networking scenario.

References

Baker, J. E. (1985). Adaptive selection methods for genetic algorithms. In *Proceedings of an
    International Conference on Genetic Algorithms and their applications* (pp. 101–111).
Bridges, C. L., & Goldberg, D. E. (1987). An analysis of reproduction and crossover in a
    binary-coded genetic algorithm. *Grefenstette, 878*, 9–13.
Brindle, A. (1981). Genetic algorithms for function optimization (Doctoral dissertation and
    technical report TR81-2). Edmonton: University of Alberta, Department of Computer Science.
Chakraborty, M., & Chakraborty, U. K. (1997). An analysis of linear ranking and binary
    tournament selection in genetic algorithms. In *Information, Communications and Signal
    Processing, 1997. ICICS., Proceedings of 1997 International Conference on IEEE* (Vol. 1,
    pp. 407–411).
Goldberg, D. E., & Deb, K. (1991). A comparative analysis of selection schemes used in genetic
    algorithms. *Foundations of Genetic Algorithms, 1*, 69–93.
Holland, J. (1975). *Adaption in natural and artificial systems*. University of Michigan Press.
Koza, J. R. (1992). *Genetic programming: On the programming of computers by means of natural
    selection*, (Vol. 1). Cambridge: MIT press.
Fogel, L. J. Owens, A. J. & Walsh, M. J. (1966). *Artificial intelligence through simulated
    evolution*.
Lipowski, A., & Lipowska, D. (2012). Roulette-wheel selection via stochastic acceptance. *Physica
    A: Statistical Mechanics and its Applications, 391*(6), 2193–2196.
Melanie, M. (1999). *An introduction to genetic algorithms*. Cambridge: MIT Press.
Miller, B. L., & Goldberg, D. E. (1996). Genetic algorithms, selection schemes, and the varying
    effects of noise. *Evolutionary Computation, 4*(2), 113–131.
Pizzuti, C. (2008). Ga-net: A genetic algorithm for community detection in social networks. In
    Parallel problem solving from nature–PPSN X (pp. 1081–1090). Springer: Berlin Heidelberg.
Rechenberg, I. (1965). *Cybernetic solution path of an experimental problem*.
Riolo, R. L. (1992). Survival of the fittest bits. *Scientific American, 267*(1).
Whitley, L. D. (1989). The GENITOR algorithm and selection pressure: Why rank-based
    allocation of reproductive trials is best. In *ICGA*, (pp. 116–123).
Yao, X. W., Wang, W. L., Yang, S. H., & Cen, Y. F. (2014). Bio-inspired self-adaptive rate control
    for multi-priority data transmission over WLANs. *Computer Communications, 53*, 73–83.
Zhong, J., Hu, X., Gu, M. & Zhang, J. (2005). Comparison of performance between different selection
    strategies on simple genetic algorithms. In *Computational Intelligence for Modelling, Control and
    Automation, 2005 and International Conference on Intelligent Agents, Web Technologies and
    Internet Commerce, International Conference on IEEE* (Vol. 2, pp. 1115–1121).
Zhu, C., & Yin, G. (2009). On competitive Lotka-Volterra model in random environments.
    Journal of Mathematical Analysis and Applications, 357(1), 154–170.

Chapter 9
Bio-inspired Software-Defined Networking

Abstract Bio-inspired network systems are a field of biology and computer science. These systems have gained importance in recent past for their tremendous amount of potential in solving many challenging networking issues. Bio-inspired systems take inspiration and develop model from biological organisms or biological systems. They not only offer a complete parallelism but also mimic the characteristics, laws, and dynamics between biological systems and network systems. Biological organisms have self-organizing and self-healing characteristics that help them in achieving complex tasks with much ease. Software-defined networking (SDN) provides a breakthrough in network transformation. It decouples the software from hardware firmware by disengaging the data plane and control plane of the networking device. Evolution of SDN in the current network scenario has enabled programmable networking that has provided a radical new way of networking. However, increasing network requirement and focus on the controller for determining the network functionality and resources allocations aims at self-management capabilities. The two systems, viz., SDN and biological organisms aim for self-organization and self-healing properties and thus, there exists a match between the two. The study provides a list of bio-inspired solutions in various issues of SDN.

9.1 Software-Defined Networking

Software-defined networking (SDN) is an emerging architecture providing services on priority basis for real-time communication. It procures in pulling out the intelligence from the hardware network to develop a management system for effective networking.

In traditional systems, the functionality performed by a switch/router was written in its firmware. A single networking device such as the router has its own hardware, intelligence, and program configuration for forwarding and controlling the path of the data packets. Furthermore, networking in 1999 was mostly concerned with the speed of file systems, i.e., how fast the transfer of files occurs from a system to

© Springer International Publishing Switzerland 2016   107
H. Rathore, *Mapping Biological Systems to Network Systems*,
DOI 10.1007/978-3-319-29782-8_9

another. In 2005, real-time communication such as VoIP and digital surveillance came into picture focusing on latency rather than the speed. Hence, the QoS was on demand. Nevertheless, occasionally FTP traffic was predominating the SIP traffic. Therefore, a better prioritization and traffic deployment management were required.

SDN separates the routing intelligence from the device itself (Jagadeesan and Krishnamachari 2014). The separation of the forwarding hardware from the control logic allows easier deployment of new protocols and applications, straightforward network visualization and management, and consolidation of various middleboxes into software control (Nunes et al. 2014). SDN also prioritizes and dynamically shapes the network traffic on demand. Additionally, it bifurcates the networking plane into data plane and control plane with services such as firewall in middle as shown in Fig. 9.1.

The data plane contains all the switches, routers, and bridges that perform packets transmission. The control plane is a set of management servers which connects to all the different networking equipments that are present in the data plane. The control plane allows the operation of a possibly disparate set of devices from a single vantage point. OpenFlow is the control protocol that controls the networking device for providing communication between the control plane and the data plane. Consider a scenario where H1 host sends a HTTP request to host H4 as shown in Fig. 9.2.

In the traditional systems, H1 would send a *SYNC* packet to the switch, where the switch looks into the routing table and sends the packet to H4. H4, on receiving the *SYNC* packet, sends the acknowledgment packet (ACK) back to H1. In the traditional switch, the IP next hop address is indicated in the routing table, and the IP-to-MAC resolution is performed at packet forwarding time using ARP. However, in case of OpenFlow controller, switch performs the forwarding of packets with the help of flow table which contains match and rule functions. In the given example, the

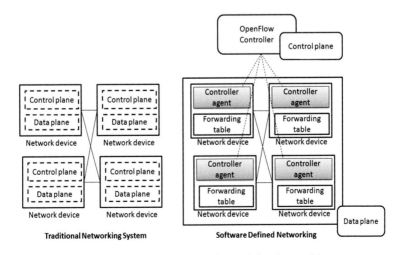

Fig. 9.1 Traditional networking systems versus software-defined networking

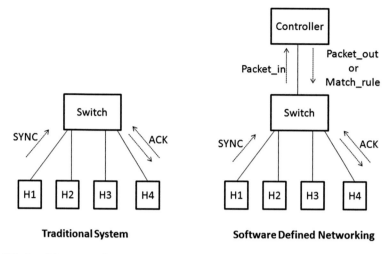

Fig. 9.2 OpenFlow controller

switch on receiving *SYNC* packet sends the *packet_in* to the controller where the controller looks into the flow table. The flow table performs matching and rule function and sends *packet_out* or *match_rule* based on the iteration of the packet. H4, on receiving the packet sends the ACK to H1 host. Thus, the difference between the traditional routers and OpenFlow is that in the OpenFlow approach, the destination MAC address of the IP next hop must be known when setting the rule.

SDN holds promising future since it provides better automation and visibility, increased uptime, and less drain in resources.

9.2 Wireless Software-Defined Networking

OpenFlow was initially developed for wired systems. In wireless networks, traffic congestion is one of the critical issues that have to be dealt. Thus, an efficient and important feature of wireless SDN is virtualization. Virtualization aims on the ability to slice the network, based on users, subnets, or traffic and allows many benefits. Flows of different slices are isolated from each other. The isolation provided by slicing enables one to run experiments safely without affecting production traffic. This ability to separate the flows space into distinct subspaces is referred to as slicing. This is made possible by inserting an application between the physical devices and the controllers, called FlowVisor. FlowVisor gives the illusion to each controller that it controls a dedicated network. Each controller acts on a slice as if it were on a dedicated network and all slices are multiplexed over the same physical infrastructure (forwarding plane) (Chaudet and Hadded 2013).

Wireless SDN has its own challenges and opportunities. It helps in increasing end-user connectivity and QoS. Additionally, it also provides efficient channel allocation which in turn helps in enhancing access point's cooperation. The wireless SDN provides better security and also localization. However, there are challenges such as slicing, channel allocation, and handoffs which are to be looked upon.

Wireless mesh network consists of routers organized in mesh topology. Mesh topology provides reliability and offers redundancy, i.e., it allows switching if the connected router stops functioning. The deployment of OpenFlow in wireless mesh network helps in switching the data to another controller when the connecting controller stops working. One of the wireless mesh network architectures is proposed in Detti et al. (2013). Here, the authors provide an OLSR to OpenFlow protocol for controlling the packets.

9.3 Security in Software-Defined Networking

Centralized controller paradigm and the ability to control the network by software prone SDN to threat. Thus, security plays a vital role in SDN. Security threats can be added in the network via various means (Kreutz et al. 2013):

Denial of Service (DoS) Attack: As SDN is an open network, the attacker can compromise any switch/router/user to congest the traffic of the network. DoS attacks aim in sending unnecessary/redundant request to congest the production traffic. A possible solution is a development of intrusion detection system (IDS).

Attack on Switch: Any compromised switch can either slow down the traffic or drop the packets in the network. This would affect the overall efficiency of the network. A trust management protocol, i.e., assignment of trust ratings to the switch can look over attacks on switch.

Attack on Control Plane Communication: Transport layer security and secure socket layers (TLS/SSL) are the cryptographic techniques used for the transport of information between the data plane and control plane. However, they do not guarantee efficient and secure communication as stated in Holz et al. (2012), Georgiev et al. (2012). Once an attacker gains access to the control plane, it may be capable of aggregating enough power force to launch DDoS attacks or even blackhole attack. Introduction of better cryptographic techniques can overcome the attack on control plane communication. Threshold cryptography across controller replicas (where the switch will need at least n shares to get a valid controller message) is one such solution (Desmedt 1994).

Attack on Controller: Fraudulent controllers can compromise the whole network and introduce unexpected behavior. Replication of data is one of the solutions for compromised controllers.

Attack between the Controller and Management Applications: This attack is analogous to control plane communication attack. Malicious applications can now be easily developed and deployed on controllers by means of this attack. Better trust management and cryptographic techniques provide a promising solution.

Attacks on Administration Stations: Administrative stations were used to monitor controllers. However, compromising the administrative stations can take hold of the entire network. Use of double certification, i.e., requiring the credentials of two different users to access a control server yields a solution for the mentioned attack.

Lack of Trustworthiness: In order to investigate about an attack, reliable information from all components and domains of the network is required. The authenticity of the information provided by them should be ensured. Logging and tracing are techniques used for checking the trustworthiness of the information provided by components of the network.

Among the list of attacks mentioned above, attack on controllers is the major one. Since the controllers are the main component of SDN, a secure and dependable controller provides better resilience to the entire network from the attacks. A better and efficient trust management with self-healing mechanism to recover from the attack provides a feasible and efficient solution for SDN controllers.

9.4   Bio-inspired Solutions for Software-Defined Networking

9.4.1   Self-organization and Stability in Software-Defined Networks

Today, networking systems require self-organizing capabilities. If OpenFlow controllers in SDN perform load balancing and traffic engineering, it will not only improve network performances but also can ease human operators to work and postpone resources investments (Manzalini 2013). The controllers' self-organization and stability thus play a vital role in SDN. Ecosystem can work as an inspiring model for providing the self-organization characteristic in SDN.

Ecosystem is a class of species trying to adapt in changing environmental conditions. Self-organization of ecosystems happens to ensure stability in network population. Consider for instance ants, they interact with each other, and with the environment, thus cross-influencing their behaviors and having an impact on the environment to ensure stability and optimization.

The work presented by Manzalini (2013) argues that self-organization can be adapted in SDN through Kelly's network optimization rules for traffic control (Kelly 1997; Anastasi et al. 2005). The ecosystem utility function was adopted to maximize the profit with minimum cost. The proposed methodology initially decomposes the network components followed by performing utility function on each controller which is then aggregated to have cumulative utility function for all the controllers. It allows offloading to traffic load of one controller to another controller if one becomes dysfunctional/overload. Different controllers can either complement or compensate each other with the help of the following function:

$$U(x_1, x_2, \ldots x_n) = \frac{\prod_i [1 + Kk_i U_i(x_i)] - 1}{K} \qquad (9.1)$$

where K is a normalization constant ensuring that the utility values are scaled over the range space between 0 and 1.

9.4.2   Fault Management in Software-Defined Networking

There are two well-known failure recovery mechanisms in OpenFlow networks, restoration and protection. Restoration matches well with the OpenFlow principle that a network is controlled by a centralized controller. If a failure occurs, the controller calculates alternative paths for every affected flow and sets up new flow entries to switches along the new path. Therefore, the time taken to recover from a failure with the restoration mechanism is proportional to the number of affected flows and the length of the new path. Protection works on the principle of group table. The group table consists of group entries that contain action buckets. When a packet arrives at a switch, a matching flow entry is examined first. If there is a corresponding group entry, the packet is redirected to the corresponding group entry. If there is failure, the switch would pass the packet to the next alternative path.

9.4.3   Cognition: A Tool for Reinforcing Security in Software-Defined Network

Security plays a vital role in the networks since it ensures reliability to the network. A cognition-based framework is proposed in Tantar et al. (2014). According to it, the classification techniques can build up tools for detecting the anomalies and cognitive algorithms can be used in learning topology and anticipating failures. The implementation is done on the application layer that is attached to the controller through northbound interface as shown in Fig. 9.3.

The control plane receives the knowledge and the application layer performs the cognitive reasoning processing in the network. A cognitive reasoning with three strategic capabilities are maintained, viz., strategic(deliberate), tactic(reflexive), and reactive. Strategic capability anticipates using heuristic approaches for learning. Tactic capability performs profiling to filter data and data dimensionality reduction. They supervise the interaction between the strategic and reactive processes.

Application plane

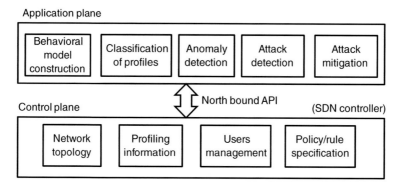

Fig. 9.3 Schematic for cognition-based security in SDN (*Source* Tantar et al. 2014)

Reactive capability uses feedback control loops for the prevention and mitigation of anomalies. They take immediate responses based upon the reception of an appropriate external stimulus.

9.4.4   Control Loops for Autonomic Systems

Control loops are another way to deal with fault management in SDN. Control loops for autonomic systems work via adding sensors and effectors over the data layer which orient, observe, decide, and act accordingly. These processes occur in parallel as shown in Fig. 9.4.

Cognitive network management architecture was further proposed having a hybrid approach of cognitive control loops for fault management in SDN (Kim 2013).

9.4.5   Self-governance and Self-organization in Autonomic Networks

Autonomic capabilities can be achieved through self-knowledge, self-governance, and self-organization. Self-knowledge provides the ability to learn from experience,

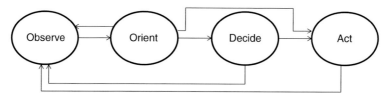

Fig. 9.4 Control loops for security in SDN (*Source* Kim 2013)

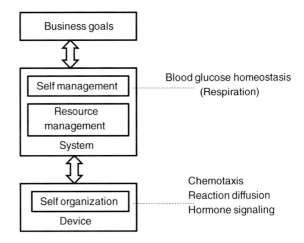

Fig. 9.5 Biological systems to autonomic systems (*Source* Balasubramaniam et al. 2006)

self-organization ensures coordination by local interactions, and self-governance aims at achieving business goals. The three capabilities are achieved taking inspirations from field microbiology as shown in Fig. 9.5:

- *Use of blood glucose homeostasis techniques to achieve overall system resource management*: The blood glucose homeostasis is one basic requirement to maintain system equilibrium. Glycogen and fat production in the body are maintained through this process.
- *Use of reaction–diffusion-like principles to achieve decentralized self-organization*: Reaction–diffusion techniques are used in the body for tissue development.
- *Use of chemotaxis for route discovery*: Chemotaxis is the ability to move based on stimulus attraction which supports the ability of self adaptation.
- *Use of hormone signaling for discovering neighbor resources and services*: Cell-to-cell communication is required in blood stream for transmitting the chemical messages.

Blood glucose homeostasis is used at the control layer for providing load and traffic management. The rest of the methods are used at the device level for maintaining link failures, addition of new devices, etc.

9.5 Summary

This chapter provides insights on SDN which is one of the emerging technologies of networking. In SDN, self-configuration, self-healing, self-optimization, and self-protection are necessary requirements and looking at the analogies from biological systems, they seem to act perfect to attain these characteristics. This chapter provides details on how biology has helped in solving issues of software-defined networking.

References

Anastasi, G., Borgia, E., Conti, M., & Gregori, E. (2005). Rate control in communication networks: Shadow prices proportional fairness and stability. *Journal of Cluster Computer, 8*(2–3), 135–145.

Balasubramaniam, S., Botvich, D., Donnelly, W., Foghlú, M. Ó., & Strassner, J. (2006). Biologically inspired self-governance and self-organisation for autonomic networks. In *Proceedings of the 1st International Conference on Bio Inspired Models of Network, Information and Computing Systems* (p. 30). ACM.

Chaudet, C., & Haddad, Y. (2013). Wireless software defined networks: Challenges and opportunities. In *2013 IEEE International Conference on Microwaves, Communications, Antennas and Electronics Systems (COMCAS)* (pp. 1–5). IEEE.

Desmedt, Y. G. (1994). Threshold cryptography. *European Transactions on Telecommunications, 5*(4), 449–458.

Detti, A., Pisa, C., Salsano, S., & Blefari-Melazzi, N. (2013). Wireless mesh software defined networks (wmSDN). In *2013 IEEE 9th International Conference on Wireless and Mobile Computing, Networking and Communications (WiMob)* (pp. 89–95). IEEE.

Georgiev, M., Iyengar, S., Jana, S., Anubhai, R., Boneh, D., & Shmatikov, V. (2012). The most dangerous code in the world: validating SSL certificates in non-browser software. In *Proceedings of the 2012 ACM Conference on Computer and Communications Security* (pp. 38–49). ACM.

Holz, R., Riedmaier, T., Kammenhuber, N., & Carle, G. (2012). X. 509 Forensics: Detecting and localising the SSL/TLS Men-in-the-middle. In *Computer Security–ESORICS 2012* (pp. 217–234). Berlin: Springer.

Jagadeesan, N. A., & Krishnamachari, B. (2014). Software-defined networking paradigms in wireless networks: A survey. *ACM Computing Surveys (CSUR), 47*(2), 27.

Kelly, F. (1997). Charging and rate control for elastic traffic. *European Transactions on Telecommunications, 8*(1), 33–37.

Kim, S. (2013). Cognitive Model-Based Autonomic Fault Management in SDN (Doctoral dissertation, Ph. D. thesis, Pohang University of Science and Technology).

Kreutz, D., Ramos, F., & Verissimo, P. (2013). Towards secure and dependable software-defined networks. In *Proceedings of the Second ACM SIGCOMM Workshop on Hot Topics in Software Defined Networking* (pp. 55–60). ACM.

Manzalini, A. (2013). Self-Organization and stability in software networks. *International Journal of Information Engineering, 3*(2), 30–36.

Nunes, B., Mendonca, M., Nguyen, X. N., Obraczka, K., & Turletti, T. (2014). A survey of software-defined networking: Past, present, and future of programmable networks. *Communications Surveys & Tutorials, IEEE, 16*(3), 1617–1634.

Tantar, E., Palattella, M. R., Avanesov, T., Kantor, M., & Engel, T. (2014). Cognition: A tool for reinforcing security in software defined networks. In *EVOLVE-A Bridge between Probability, Set Oriented Numerics, and Evolutionary Computation V* (pp. 61–78). Springer International Publishing.

Chapter 10
Case Study: A Review of Security Challenges, Attacks and Trust and Reputation Models in Wireless Sensor Networks

Abstract In wireless sensor network (WSN), where nodes besides having its inbuilt capability of sensing, processing, and communicating data, also possess some risks. These risks expose them to attacks and bring in many security challenges. Therefore, it is imperative to have a secure system where there is perfect confidentiality and correctness to the data being sent from one node to another. Cooperation among the nodes is needed so that they could confidently rely on other nodes and send the data faithfully. However, owing to certain hardware and software faults, nodes can behave fraudulently and send fraudulent information. Nevertheless, since the network is openly accessible, anybody can access the deployment area which breaches the security of WSN. Therefore, it is required to have correct and accurate secure model for WSN to protect the information and resources from attacks and misbehavior. Many researchers are engaged in developing innovative design paradigms to address such nodes by developing key management protocols, secure routing mechanisms, and trust management systems. Key management protocols and secure routing cannot itself provide security to WSNs for various attacks. Trust management system can improve the security of WSN. The case study begins by explaining the security issues and challenges in WSN. It discusses the goals, threat models, and attacks followed by the security measures that can be implemented in detection of attacks. Here, various types of trust and reputation models are also reviewed. The intent of this case study is to investigate the security-related issues and challenges in wireless sensor networks and methodologies used to overcome them. Furthermore, the present case study provides details on how bio-inspired approaches in WSN prove a benefactor in many ways.

10.1 Introduction

Security issues have been studied on a variety of networks such as, wired networks, wireless networks, social networks, etc. It is an important aspect which promises not only the reliability to the system, but also helps in sustaining the network. It is

© Springer International Publishing Switzerland 2016 117
H. Rathore, *Mapping Biological Systems to Network Systems*,
DOI 10.1007/978-3-319-29782-8_10

imperative to have a secure system where there is perfect confidentiality and correctness to the data being sent from one system to another.

WSN often monitors critical infrastructure and carries sensitive information which makes them desirable targets for attacks. Furthermore, since the network is openly accessible, anybody can access the deployment area which breaches the security of WSN. Attacks may be facilitated by Dargie and Poellabauer (2010):

(1) *Remote and unattended operation.*
(2) *Wireless communication.*
(3) *Lack of advanced security features due to cost, form factor, or energy.*

As a consequence, sensor networks require effective solutions to overcome from various types of attacks. Nevertheless, while designing efficient solutions for WSN security various challenges have been addressed.

10.2   Constraints on Wireless Sensor Networks

Obstacles are observed while setting up a WSN (Ringwald and Romer 2007). Resources of individual sensor nodes are intrinsically constrained in a WSN. They have limitations such as inadequate processing capability, communication bandwidth, and storage capacity. The causes of having such limitations are due to limited energy and physical size. Additionally, these small-sized nodes are highly expensive. Nevertheless, the processor and the radio used in sensor nodes consume a lot of energy while sending, processing, and receiving data. Major cause of energy consumption includes sensor transducer, communication hardware, and microprocessor computation. Owing to this, optimum communication with minimal processing is recommended in sensor nodes (Soderman 2008).

Furthermore, as sensor nodes are deployed in hostile conditions where access to permanent power supply is negligible, nodes are configured with a battery source. Over and above that, if the battery becomes deficient, nodes turn counterproductive. Also, since the nodes are remotely located, anybody can hamper the node not only at data level but also at infrastructure level, to breach the security of the network. As a result, it is critical to secure the overall system.

10.3   Threat Model

In WSNs, it is a common practice to assume that the attacker might be aware of the security measures that are deployed in a sensor network (Wang et al. 2006). They may be able to compromise a node or even physically capture a node which in turn allows the attacker to steal the information within the node. Attacks in WSN can be classified into:

(1) *Outsider versus insider attacks*: The cause of an outsider attack is due to nodes which do not belong to the network. The insider attacks cause nodes of the same network to behave in an unintended manner.
(2) *Passive versus active attacks*: Passive attacks comprise eavesdropping or monitoring packets exchanged within a WSN; active attacks involve some modifications of the data steam or the creation of a false stream.

The metrics that can be used to evaluate security scheme in WSN should make the system resilient, energy efficient, flexible, and scalable. They should be fault tolerant, should have self-healing capability and high assurance.

10.4   Security Requirements for Wireless Sensor Networks

The goals of security services in WSN are aimed to protect the information and resource from attacks and behavior (Sharma et al. 2010). The basic goals are a set of properties which are listed below (Lee 2007):

(1) *Confidentiality*: The goal of confidentiality is to prevent any unauthorized usage or revelation of information in transit.
(2) *Integrity*: It aims to ensure that data being transferred has not been changed by the attacker. It safeguards the accuracy and completeness of the information.
(3) *Availability*: It is the quality of being at hand when needed which states that authorized users should have reliable and timely access to information.
(4) *Access control*: This property is required to permit or deny the use of an object by a subject.
(5) *Authentication*: It aims at validating a claimed identity of an end user or a device.
(6) *Authorization*: It is an act of granting access rights to the user.
(7) *Accountability*: It is the act of being explained and justifiable.
(8) *Freshness*: Data should be fresh and nonredundant.
(9) *Robustness*: It ensures the network to be functional as well as capable of handling errors in catastrophic conditions and environment.

Consequently, while designing a security management system for the network, all the security properties should be attained for a concrete solution.

10.5   Focus and Contents

The present chapter gives the background for a secured WSN. The chapter is broadly organized into five sections namely attacks, key management protocol and trust-reputation model, recovery using bio-inspired model, and intelligent water drops as shown in Fig. 10.1. Section 10.6 throws light on various types of attacks in

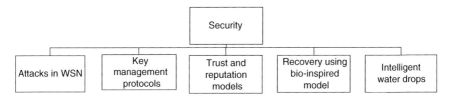

Fig. 10.1 Organization of Chapter: the chapter is divided into five sections namely attacks, key management protocol, trust-reputation models, recovery using bio-inspired model and intelligent water drops

WSN. Sections 10.7 and 10.8 illustrate ways to overcome from the attacks, where Sect. 10.7 gives a basic insight into key management protocol and Sect. 10.8 presents the details of trust and reputation models in WSN. Section 10.9 provides details on the way to recover from low trust of a node. Section 10.10 investigates on intelligent water drops scheme which can applied for effective path recovery.

10.6   Attacks in Wireless Sensor Networks

Attacks randomly come into picture which breaches the security of the system. It can aim on hindering the secrecy and authentication of the information by tampering the service integrity or it can modify and consume network availability. Attacks in WSN can be broadly classified into attack on data and attack on infrastructure (See Fig. 10.2).

10.6.1   Attacks on Data

Attacks on data can be simply defined as deviation from the expected normal behavior (Ni et al. 2009). Attacks on data focus on stealing and modifying the data which includes injecting wrong packets, deleting packets, and modifying the content. Data faults are classified on the basis of data centric and system centric views. Data centric faults change the data semantics and system centric faults are due to physical malfunction in the node. These two approaches are not disjoint and can overlap. For instance, spike fault can look like connection/hardware failure fault and stuck-at fault could look like clipping attack as shown in Fig. 10.3.

(a)   Data Centric Faults

Data centric faults can be classified on the basis of Baljak (2012):

(1)   Continuity of the occurrence
(2)   Frequency of the occurrence
(3)   Observable and learnable pattern

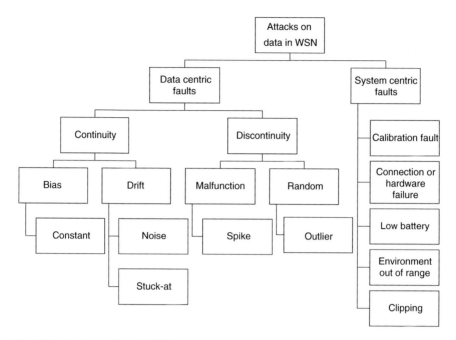

Fig. 10.2 Attacks on Data in WSN: attacks on data modifies the data content whereas attacks on infrastructure claims the identity of the nodes and consume the resources of the nodes. Attacks on data can be further classified on data centric and system centric views. The two classification approaches are not completely disjoint and can overlap. The classification is on the basis of occurrence and pattern of fault

If a sensor reading is represented with $x_i + \varepsilon_i$, where x_i is the state information which is the physical parameters, such as temperature, pressure, etc. and ε_i is a fault, fault classification can be defined on the basis of continuity or discontinuity of occurrence as follows:

Continuity of the occurrence:

In continuous occurrence of faults, the sensor returns faulty data constantly, and a pattern can be observed in the form of a function (bias or drift). In bias, the function of the error is a constant, i.e., $\varepsilon_i = constant$. The generated fault is termed as constant fault. It can have a positive or a negative offset.

In drift, the deviation of data follows a learnable function, such as a polynomial change.

It can be observed in three forms:

(1) *Noise*: High noise and variance makes sensor values experience unexpectedly high variation which can be due to hardware failures, environment out of range, or a weakening of battery supply. It is a very common and expected fault in sensor nodes.

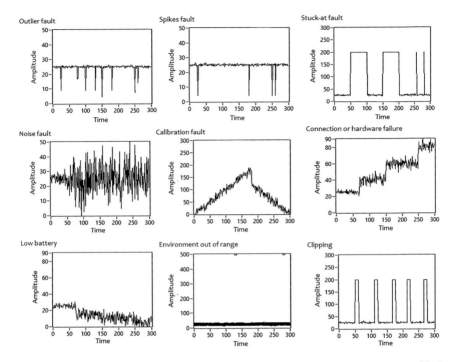

Fig. 10.3 LabVIEW Simulations on Attacks on Data in WSN: The occurrence and pattern of fault is consistent in case of noise fault, calibration fault, connection or hardware failure, low battery and clipping fault. However, it is inconsistent in case of outlier, spikes, stuck-at and environment out of range faults

(2) *Stuck-at fault*: Stuck-at is a series of data value that experiences zero variation and follows an unexpected jump in the behavior for a period of time. It is termed as stuck-at fault, since the data values get stuck-at one point and do not change their behavior for times greater than expected. The cause of this fault is due to sensor hardware malfunction.

(3) *Clipping*: A type of system centric fault which is caused when the environment exceeds the range of analog to digital convertor. It exhibits either a ceiling or a floor function. It is defined under continuous faults because it follows a learned pattern where the frequency of occurrence is fixed.

Discontinuity of the occurrence:

Discontinuous means to occur from time to time on aperiodic basis, i.e., occurrence of ε_i is discrete. It can be further divided into malfunction or random fault.

(1) *Malfunction fault*: The frequency of occurrence of faulty readings is $\varepsilon_i > \tau$, where τ is threshold frequency. Furthermore, there is no observable pattern in the fault occurrences. The fault which comes under this category is outlier faults. In case of outliers, isolated data points are observed on different time

instances. The distance from the expected behavior is high and is often caused by unknown software inserted by the attacker.

(2) *Random fault*: The occurrence of the fault is infrequent and $\varepsilon_i < \tau$. Spike fault comes under this category where multiple isolated data points are observed outside the expected range of behavior. Random faults can occur due to the connection failures or other hardware problems like battery failure.

(b) System Centric Faults

System centric faults are the ones which are observed during sensor node deployment and while understanding real-world parameters. They are of various types (Ni et al. 2009):

(1) *Calibration fault*: Calibration of transducer response is required to increase the accuracy of the sensor nodes since their factory calibration might not be well suited to the conditions of the field. There should be consistent one to one mappings of output values to input values. Calibration error can also be of three types namely offset, gain, and drift. In offset error, sensor faults are offset from the expected range by a constant amount. In gain error, measurements have offset with some gain value f which is a random multiplicative factor such as $x_i + x_i \times f$, where f lies within the interval [0.1, 10.0]. In drift errors, measurements drift away from the original values.

(2) *Connection or hardware failures*: Poor connections or hardware malfunction of sensors can cause the sensor nodes to deviate from the actual measurements. Hardware malfunction occurs due to environment perturbations or sensor age. It is categorized by sending either very low or very high sensor measurements.

(3) *Low battery*: Battery is one of the limiting factors of sensor nodes. If the battery becomes deficient, sensor turns counterproductive. Battery losses occur when processor and the radio used in sensor nodes consume a lot of energy while sending, processing, and receiving data. System performance is affected by battery supply by introducing noise and inaccurate measurements depending upon the type of sensor.

(4) *Environment out of range*: When the sensitivity range of transducer falls outside the acceptable range, errors due to environment occur. This will give way to much higher noise or flattening of the data.

(5) *Clipping*: A type of system centric fault (explained in data centric fault).

The observable pattern of all the attacks on data can be represented pictorially as shown in Fig. 10.3. The results were generated in LabVIEW taking into account the fault pattern described above.

10.6.2   Attacks on Infrastructure

Attacks on infrastructure claim the identity of the nodes and consume the resources of the compromised nodes. The attacks on infrastructure try to utilize the resources

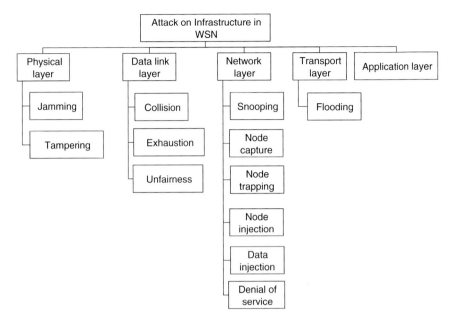

Fig. 10.4 Attacks on Infrastructure in WSN: attacks on infrastructure claims the identity of the nodes and consume the resources of the nodes. Classification is based on layer protocol of WSN i.e. physical layer, data link layer, network layer, transport layer and application layer

of the network by sending unnecessary packets to the victim. These attacks can be seen at different layers of WSN protocol (Sharma and Ghose 2010). The protocol stack in sensor nodes contains physical, data link, network, transport, and application layers and the attacks as shown in Fig. 10.4.

(a) Physical Layer

Physical layer is responsible for frequency being selected, generation of carrier frequency, modulation, data encryption, and signal deflection. At the physical layer, the attacker tries to jam the radio transmission by *jamming*. As the sensor nodes have limited power, this attack consumes a lot of energy. *Tampering* is another attack which can be seen at physical layer where the attacker compromises nodes and extracts sensitive information such as cryptographic keys or other data.

(b) Data Link Layer

Data frame detection, medium access, error control, and multiplexing of data streams are managed by data link layer. At MAC layer, the attacker can send or violate guaranteed time slots forcing the other sensors to retransmit the data multiple times for consuming power. *Collision* is one of the major attacks where two sensor nodes transmit information on the same frequency. When the packets collide, changes occur in the data portion resulting in the deviation of checksum at the receiving end. *Exhaustion* is another attack where repeated collision occurs to cause resource exhaustion. *Unfairness* is another such attack, where the attacker causes other nodes to miss their transmission deadline in a real-time MAC protocol.

(c)   Network Layer

Forwarding of packets and the assignment of addresses is performed by network layer. At network layer, sending data long ways and disturbing the routing protocol can devastate global network performance. The common examples include spoofing, altering, and replaying routing information. Blackhole, sink-hole, sybil, wormholes, hello flood, and acknowledgment spoofing are the attacks observed at the network later. Some of the network layer attacks are:

Snooping
As WSN uses radio communication for transmission, any attacker which is in the range of the network can monitor the communication. It can listen to the data and record it.

Node capture
As nodes are not placed in a secure location, attacker can capture the node and can get access to all its information. Blackhole/sinkhole attack is a type of node capturing attack (Ahmed et al. 2005). An attacker acts as a blackhole and attracts all the traffic in the sensor network. See Fig. 10.5a.

Node trapping
In node trapping attack, the attacker can trap the node and can tunnel the data to some other place as shown in Fig. 10.5b. Wormhole is a type of node trapping attack.

Node injection
To accomplish a task, the sensors may be required to work together in WSN. The sensors can accomplish the task by usage of redundant information and distribution of subtasks. If multiple identities are used by an attacker, then this type of attack is classified as Sybil attack (Fig. 10.5c). Sybil attack can be performed for attacking

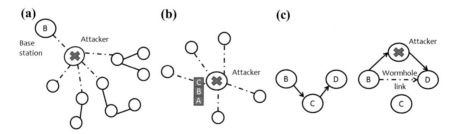

Fig. 10.5 Network Layer Attacks: **a** *Blackhole attack*: Attacker captures the relay node and attracts all the traffic in the sensor network; **b** *Wormhole attack*: B base station broadcasts routing requests messages to all the nodes in the network where the attacker receives this packet being in the range of B; **c** *Sybil attack*: Attacker claims the identity of various genuine nodes (*Source* Pathan et al. 2006)

the routing mechanism, distributed storage, data aggregation, voting, resource allocation, and misbehavior detection (Newsome et al. 2004).

Data injection
The attacker can inject fraudulent data without joining into the network. Replay attack is an example of data injection attack where the data is either delayed or repeated (Lupu 2009).

Acknowledgment spoofing
Acknowledgments are sometimes required to be transmitted by routing algorithms used in sensor networks. Here, the acknowledgments of overheard packets are spoofed by the attacker. These packets are intended for neighboring nodes and provide them with false information. A typical situation is claiming the node as alive when in fact it is dead.

(d) Transport Layer

Transport layer is responsible for specifying the reliable transport of packets. *Flooding* is one such attack where an attacker may exhaust the resource which each connection requires. In such a scenario, the attacker may also reach a maximum limit of attempted connections.

(e) Application Layer

Application layer is responsible for specifying how the data is requested and provided for both individual sensor nodes and interaction with the end user. At application layer, applications can be forced to do extensive computation which uses the limited memory of sensor nodes.

   The taxonomy of various types of faults and attacks elucidates that, there is a requirement of secure transmission of data or a mechanism where the identification of fraudulent nodes can be done which are behaving in an unpredictable manner.

10.7   Key Management in Wireless Sensor Networks

Key management protocol is used to encrypt and authenticate the data which can limit the attacks of WSN. The protocol must establish a key between all sensor nodes that wants to exchange data securely. The general idea behind this technique is to process, revoke, and distribute keys. The hardest part in key management is the establishment and revocation of keys. The steps involved in key management protocol are:

(1) At the outset, nodes are initialized. It is the process of loading initial security settings.
(2) In the second place, keys are generated in a secure and random manner.
(3) To negotiate a key between the two sensor nodes without allowing the attacker to find out the key, a key establishment protocol is used. In this step, key is

transported with the help of key agreement steps. There are three types of general key agreement schemes: *trusted-server scheme, self-enforcing scheme*, and *key pre-distribution scheme* (Zia et al. 2007). Trusted server scheme like Kerbores depends on trusted server (Neuman and Ts'o 1994). This type of scheme is not preferable in WSN because of lack of trusted server. Self-enforcing scheme uses asymmetric cryptography, such as public key certificates, Diffie–Hellman key agreement (Großschädl et al. 2007) or RSA (Kayalvizhi et al. 2010). However, because of computation and energy constraints this scheme is seldom used. Key pre-distribution is the third type of key agreement scheme where the distribution of key information is performed prior to sensor node deployment. Multiple key pre-distribution schemes are applied in WSN. Allowing all the nodes to carry a master secret key can be one of the solution. To achieve key agreement, new pairwise keys can be obtained by any pair of nodes utilizing the master secret key. Another solution is to let each sensor carry n − 1 secret pairwise keys, each of which is known only to this sensor and one of the other n − 1 sensors (assuming n is the total number of sensors) (Du et al. 2006). The security of communications among nodes is not affected while compromising a single node. This provides an advantage to the scheme and makes it more resilient. However, as *n* could be large, this scheme is impractical for sensors which have limited amount of memory.

(4) A distributed system is required to control the keys for minimizing the communication. To avoid transporting large amount of usage information, the system is divided into key classes.

(5) Later, revocation of broken keys is done. Revocation can be done using a trusted third party or performing voting in the network or via time stamps. Key revocation scheme was initially introduced by Eschenauer and Gligor (2002). Here, the controller node broadcasts message to each node in the network informing about the compromised keys. A signature key is sent to each node prior to the broadcast message. Schemes were introduced later where revocation was made using a trusted third party or performing voting in the network or via time stamps (Soderman 2008).

(6) Moreover, keys are to be stored. Keys should be minimal because of memory constraints of sensor nodes. Therefore, if the attacker gets physical access to the sensor node, its keys are broken to release memory.

The computation requirement for developing keys is very high. This is because; the embedded processors used in wired or ad hoc networks are more powerful than those used in WSN. As such, complex cryptographic keys cannot be used in WSNs. Also, the memory used in sensor nodes usually includes flash memory and RAM. Flash memory is used for storing information and RAM is used for processing the computation. Therefore, there is not enough space to run complicated algorithms in sensor nodes. Also, a secure way for sending the private key is a challenge when you have large number of nodes. Key management protocols cannot alone provide security to WSNs for the attacks described in the above section. Therefore, WSN should have mechanisms to detect and react to various intrusions. An intrusion detection method could be a better solution in this respect where the network monitors for suspicious activity outside the normal and expected behavior. For this, cooperation among sensors is required to detect and nullify the effect of intruders in the network. Cooperation can be attained by building trust and reputation models for WSN.

10.8   Trust and Reputation Model

Trust and reputation are very important factors in decision-making processes of any network where uncertainty is a major factor for ruining the overall system. *Trust is a personal and subjective phenomenon that is based on various factors or evidence* (Jøsang 2007) and *reputation is a collective measure of trustworthiness in a sense of reliability based on the ratings from the members of a community* (Boukerch et al. 2007).

In the context of wireless sensor network, while sensing data, nodes can produce ratings based on direct observation (known as first-hand information) and indirect observation (also known as second-hand information) as shown in Fig. 10.6. Direct observation is the trust ratings calculated selflessly, by taking into account many factors, such as packet delivery ratio, amount of energy left, hardware error, deviation from sensor readings, etc. It is more or less based on observations and experience, whereas indirect observation contains the observations from other

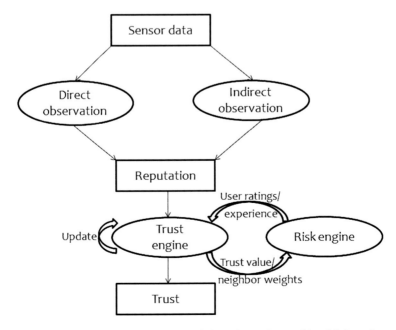

Fig. 10.6 Trust calculation based on first-hand information and second-hand Information

nodes. Based on first-hand information and second-hand information, trust is calculated. Trust engine regularly calculates trust ratings based on risk engine decision (Momani and Challa 2010). Both trust engine and risk engine are dependent on each other in terms of user ratings and neighbor weights. Finally, trust ratings are generated which are regularly updated with time.

10.8.1   Methods to Calculate Trust

There are four basic models to calculate trust as shown in Fig. 10.7 (Reddy 2012):

(1) *Reputation-Based Model*: Uses packet transfer rate to calculate trust.
(2) *Event-Based Model*: Trust is calculated at particular or specific timed events or periodically.
(3) *Collaborative Model*: Models are developed to calculate trust based on first-hand and second-hand information.
(4) *Agent-Based Model*: In this model, an agent node is introduced from a cluster of nodes within communication distance. This agent node is used to store the packet transfer information. Initially, all the nodes are given r-certificate which shows either they are highly trusted or not trusted at all (Mármol and Pérez 2011). Now consider two nodes n_i and n_j which want to communicate with each

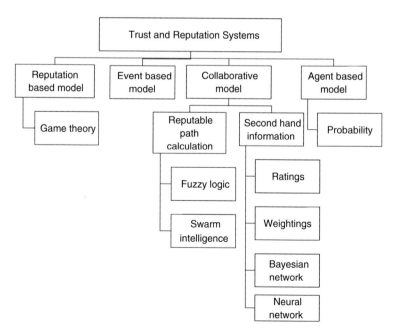

Fig. 10.7 Trust and reputation models classification: The classification is on the approaches, such as reputation based, event based, collaborative based and agent based

other. Here, n_i asks for r-certificate of n_j. Based on the certificate provided, n_i decides whether to send information or not. After the transaction is over, n_i makes trust evaluation on n_j based on quality of service it gets and submits its evaluation to its local mobile agent and sends t-instrument (local to n_j) for n_j. On receiving it, n_j updates its r-certificate.

10.8.2 Methodologies to Model Trust

In general, trust model takes into account both direct and indirect observations. There are various models which are developed to model trust. These are listed below:

(a) Ratings

This was the initial model which was developed that calculated trust on direct summation of first-hand information and second-hand information as seen from equation:

$$T_{ij} = (T_{ij})_D + (T_{ij})_{ID} \qquad (10.1)$$

where T_{ij} is the trust of j node with respect to i node, D is the direct information and ID is the indirect information.

It was a very basic model which had less detection rate of fraudulent nodes. The model was simple to implement and did not incorporate the detection of various attacks (Resnick et al. 2000; Jonker and Treur 1999).

(b) Weightings

In this model, weights were assigned to the indirect observation (Momani 2008). Nodes monitor the behavior of other nodes and compute the trust with the help of following equation:

$$T_{ij} = (T_{ij})_D + w \times (T_{ij})_{ID} \qquad (10.2)$$

where w is the weight whose value vary between 0 and 1.

A weighted approach was proposed where initially every node was highly trusted with weights (=1) (Atakli et al. 2008). The network was adapted in the architecture between a group of sensor nodes and their forwarding node (FN) as shown in Fig. 10.8.

The FN collects all information provided by sensor nodes and calculates an aggregation result A_r using the weights w_n assigned to each sensor node as given in the formula:

$$A_r = \sum_{i=1}^{n} w_n \times x_n \qquad (10.3)$$

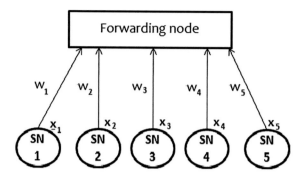

Fig. 10.8 Topology of Weightings Method: Forwarding node collects all information provided by sensor nodes SN_1 ... SN_5 and calculates an aggregation result using W_1 ... W_5 and x_1 ... x_5 (*Source* Atakli et al. 2008)

where x_n is the data acquired from sensor nodes. With the help of following equation weights w_n were updated for each node based upon the variation of A_r with respect to x:

$$w_n = \begin{cases} w_n - \theta_n & \text{if } x_n \neq A_r \\ w_n & \text{elsewise} \end{cases} \tag{10.4}$$

where θ is a constant value and r is the ratio of sensor node sending fraudulent data to the total number of nodes under the same FN.

Simulations are carried out for a network size of five nodes where initially every node has a weight equal to one and θ value equal to 0.1. Measurements of sensor nodes are in the range [10–20]. Deviation of each sensor node is checked with the mean value of past observations. If the variation was greater than 0.1 then weights are updated with the help of Eq. 10.4. Figure 10.9 shows the result depicting the rate of change of weights/trust ratings. The figure represents how the trust ratings are decreased with respect to time. The figure represents how the trust ratings are decreased with respect to time. The trust ratings decrease to zero since there is no rule for increase in trust ratings when the sensor nodes start behaving well. Once the trust ratings decrease, it never increases.

Weighted approach is found effective only for small-sized networks. For large-sized networks, number of forwarding nodes was substantially increased. However, increasing the number of forwarding nodes subjects the network to another problem known as node-clustering problem. Furthermore, the approach is based on the assumption that base stations are trusted. Hence, if the attacker could gain control over the base stations, it can do any possible attack against the WSN.

(c) Probabilistic Method

An agent-based trust model for WSN was presented in (Chen et al. 2007) to observe the behavior of nodes and broadcast their trust ratings. The agent nodes monitor the

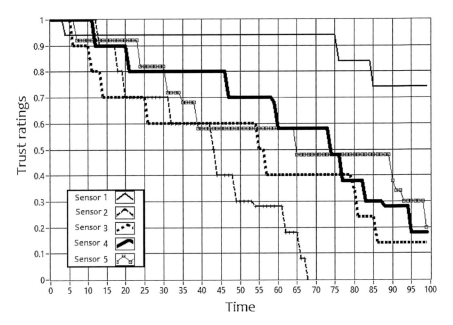

Fig. 10.9 LabVIEW Implementation of Weightings Method: Showing how the weights are decreased on account of the information provided by the neighboring nodes

sensor nodes and send the trust ratings to them. The sensor nodes on receiving them, computes and broadcasts those trust ratings. According to the received information, sensor nodes make the decision about cooperation with their neighbors. This method of calculating the trust ratings maintains state count for every positive and negative outcome/decision (Chen et al. 2008). Data collected is checked with respect to two parameters:

(1) *Routing*: which looks at the data forwarding behavior.
(2) *Data processing*: which monitors raw sensing data, performs data aggregation, and checks delay in data transmission.

For every positive outcome on above parameters, positive count (p_c) was incremented and for every negative outcome negative count (n_c) was incremented. Given $\{p_c, n_c\}$ the probability x of obtaining a positive outcome was computed as follows (Chen et al. 2007):

$$P(x) = P(x| <p_c, n_c >) = \frac{x^{p_c}(1-x)^{n_c}}{\int_0^1 x^{p_c}(1-x)^{n_c}dx} \tag{10.5}$$

Furthermore, certainty of each event was calculated using:

$$c(p_c, n_c) = \frac{1}{2} \int_0^1 \left| \frac{x^{p_c}(1-x)^{n_c}}{\int_0^1 x^{p_c}(1-x)^{n_c} dx} - 1 \right| dx \qquad (10.6)$$

Trust was then calculated using the following equation:

$$\text{Trust} = \frac{p_c + 1}{p_c + n_c + 1} \qquad (10.7)$$

Probabilistic method does not require second-hand information. It also guarantees better results than cryptographic results. However, this scheme limits attacks, such as bad-mouthing attack, on-off attack, and conflicting behavior attack.

(d) Bayesian Network

Bayesian network uses Baye's rule as the principle rule. Baye's rule expresses how a subjective degree of belief should randomly change on account for evidence (Momani et al. 2008). The advantage of this technique is that the framework is generic and allows more components to be added to and/or deleted from the framework very smoothly and transparently besides the direct and indirect observation as seen in Fig. 10.10. Using Bayes' theorem, the probability of the total trust, given the direct and indirect observation, can be presented, as shown in equation (Momani et al. 2008):

$$P(T|D, I) = \frac{P(D|T, I) * P(T|I)}{P(D|I)} \qquad (10.8)$$

A reputation-based trust system was proposed to detect the selfish nodes using Bayesian formulation specifically beta reputation system (Ganeriwal et al. 2008).

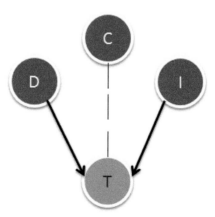

Fig. 10.10 Bayesian Network: D and I are direct and indirect observation respectively. T stands for total trust computed on account of both D and I. Framework is generic since a third component C (communication observation) can be added

In this method, reputation is calculated using a watch dog mechanism. After the packet gets passed to other nodes, the packet gets eavesdropped by each node in the network. This is performed to confirm that the packet has reached the destination. Based on the reputation scheme, the trust value is formulated.

The main disadvantage of Bayesian network technique is that the trust evaluation was based on node's QoS property which makes the false positive rates to range between 0.31 and 0.68 while the false negative rates range between 0.11 and 0.28 (Meng et al. 2013). Furthermore, flat wireless structure was used which does not allow more nodes to be added in the network (Raj et al. 2013).

(e) Game Theory

In the branch of mathematics, game theory is a method that deals with multi-person decision-making situations. It is designed to address situations in which the outcomes of a persons' decision depends not just on how they choose among the several available options, but also on the choices made by the people with whom they interact. It is used to determine a preferred strategy where such interactions are in play. A game comprises of a set of players, a set of strategies for each player, and a set of corresponding utility functions. Throughout the game, a strategy for a player is to perform complete plan of actions, in all possible situations. In any game, the players try to act greedily, according to their preferences, to maximize their consequences. A utility function expresses these preferences. These preferences map every consequence to a real number. Nash equilibrium provides a solution to describe a steady-state condition of the game. A strategy is not changed by a player unless a better strategy is available that can result in more utility that is favorable for the current player. The normal form of a game is given by a tuple (Shigen et al. 2011):

$$G = (I, S, U) \tag{10.9}$$

where G is the particular game, I is the set of players, S is the set of strategies used by the players, and U is the set of utility functions that the player wishes to maximize.

Game theory can be used to capture the nature of conflict in WSNs security. In WSN, game theoretical model tries to mathematically capture the behavior of nodes in situations where the decisions depend upon the behavior of the other nodes. Transaction between the nodes demands some resources such as energy and storage for proper functionality (Chen et al. 2011). Thus, the nodes may choose the selfish noncooperative behavior to think about the limited resources for themselves. The node can behave cooperatively (*CP*) or noncooperatively (*NCP*) giving it a game theoretical approach (Chen et al. 2011).

$$
\begin{array}{c}
\quad\quad CP \quad NCP \\
\begin{array}{c} CP \\ NCP \end{array}
\begin{pmatrix} R_p - C_e & -C_e \\ R_p & C_e \end{pmatrix}
\end{array}
\tag{10.10}
$$

Following scenario is observed:

(1) If both nodes choose cooperative strategy then they would gain R_p units of pay-off and consume C_e units of energy.
(2) If both of them use a noncooperative strategy each of them gains a zero unit of pay-off.
(3) If one chooses a cooperative and other chooses a noncooperative strategy, cooperative will gain zero and consume C_e units of energy and noncooperative will gain R_p units of pay-off.

According to different applications, a taxonomy was proposed by Shigen et al. (2011), which classifies the current existing typical game-theoretic approaches for WSNs security into four sets: preventing denial of services (DoS) attacks, intrusion detection, strengthening security, and coexistence with fraudulent sensor nodes as shown in Fig. 10.11. To prevent DoS attacks, cooperative games (Agah et al. 2004), noncooperative games (Agah et al. 2005) and repeated games (Agah and Das 2007) have been proposed. Intrusion detection includes noncooperative games (Reddy 2009) and Markov games (Alpcan and Basar 2006). To further strengthen security, auction theory (Agah et al. 2006) and coalitional games (Li and Lyu 2008) have been suggested. For the detection of fraudulent sensor nodes, signaling games have been recommended (Wang et al. 2009). Interested readers can get better insight by going through these papers.

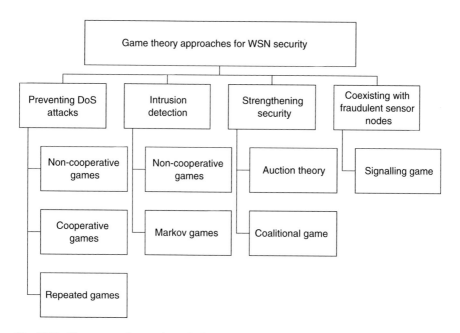

Fig. 10.11 Taxonomy of game theoretical approaches in WSN security

The main disadvantage of game theoretical technique in WSN security is its computational complexity which is hard to implement. Also, the detection percentage is ranged between 30 and 60 % (Shigen et al. 2011).

(f) Neural Network

Neural network approach employs the analytical redundancy to estimate the values provided by a sensor node by considering the past/present values given by the adjacent neighboring sensor nodes. This estimate is compared with the actual value of the sensor node to increase/decrease its trust factor accordingly. In other words, sensor node compares the present value with the estimated value to increase/decrease the trust factor accordingly where the value provided by neighborhood nodes is cumulative estimation based on present values and past history as shown in Fig. 10.11 (Curiac et al. 2007). Consider the selected sensor node numbered as 5 in Fig. 10.12 which sends the data x_A, whose neighbors 1, 2, 3, 4, 6, 7, 8, 9 send their data to the estimation and prediction block to compute x'_A. Error is calculated between the two values as:

$$Error = x_A - x'_A \tag{10.11}$$

Lesser the error between the two, more is the trust ratings and vice versa.

A typical ANN block has a block diagram as shown in Fig. 10.13a. Here, 'input' is the sensor measurement and 'desired output' is the mean of past values. It is a

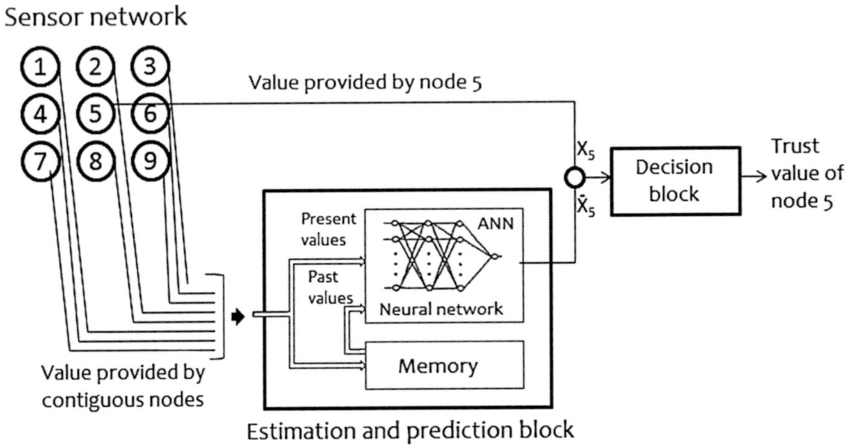

Fig. 10.12 Neural network Approach: It has two blocks namely estimation and prediction block and decision block. The estimation and prediction block has a neural network block termed as Artificial Neural Network (ANN) which takes in the present and the past values of sensor nodes (*Source* Curiac et al. 2007)

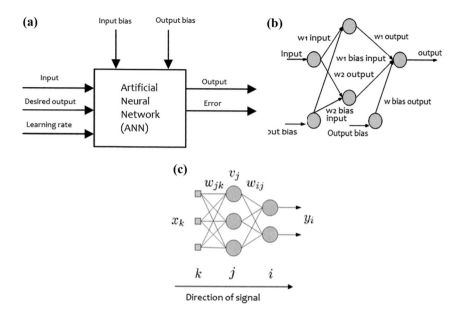

Fig. 10.13 *ANN Block*: **a** It has 'input' as the sensor measurement and 'desired output' as the mean of past values; **b** Typical 2 layer feed forward network; **c** Feed forward process direction (*Source* Auralius 2009)

typical 2-layer feed forward neural network (see Fig. 10.13b) (Auralius 2009). The activation function of the input layer is an identity function and output layer has a sigmoid function. It has two basic processes namely feedforward process and back propagation process (Jerome et al. 2005).

In the feedforward process, the network will receive single input and produce single output (see Fig. 10.13c). The input signal will be broadcasted to each neuron in the next layer with the help of following equations (Bhanot 2008):

$$h_j = \sum_k w_{jk} x_k \tag{10.12}$$

$$v_j = \frac{1}{1 + \exp(-h_j)} \tag{10.13}$$

$$s_i = \sum_i w_{ij} v_j \tag{10.14}$$

$$y_i = \frac{1}{1 + \exp(-s_i)} \tag{10.15}$$

In the back propagation process, weights are adjusted to minimize the errors with the following equations (Bhanot 2008):

$$\delta_i = y_i(1 - y_i)\left(y_i^d - y_i\right) \tag{10.16}$$

$$\delta_j = v_j(1 - v_j)\sum_i \delta_i w_{ij} \tag{10.17}$$

$$w_{ij}(t+1) = w_{ij}(t) + \eta\delta_i v_j \tag{10.18}$$

$$w_{jk}(t+1) = w_{jk}(t) + \eta\delta_j w_{ij} \tag{10.19}$$

Back propagation of errors was followed by training the neural network. Through this training, the sensor dynamics is identified by the neural network. The objective was to achieve identical behavior between the selected sensor network and the neural network. Accordingly, for the selected sensor node, current value is given as the input and the mean of past values will be given as the desired output. While on training it was necessary to adjust the bias to reduce the error. Bias was adjusted using trial and error method where first bias was set close to zero. When the neural network converges to a certain value with an intolerable error, the bias value was increased gradually. After training, the best results were achieved with the parameters shown in Table 10.1. With these values, the neural network controller shows a good performance in term of input tracking capability as seen in Fig. 10.14. The graph shows the black line as the current measurement and red line as the learned value from ANN. The lesser the gap between the two, higher are the trust ratings. The overall algorithm for the neural network approach in WSN can be explained in Algorithm 10.1.

The main disadvantage of using neural network approach in WSN, is the complexity associated with it which makes it hard to implement. The space complexity of neural network in WSN is $O(RT)$/cycle where R is the total number of neurons and T is the maximum number of activation changes (Soliman et al. 2012).

Table 10.1 Parameter values for neural network approach

Parameters	Values
$w1$ input	−3.12317
$w2$ input	−7.0356
$w1$ bias input	−0.211902
$w2$ bias input	−0.211902
$w1$ output	−3.12317
$w2$ output	−7.0356
w bias output	11.846
Input bias	0
Output bias	0.2

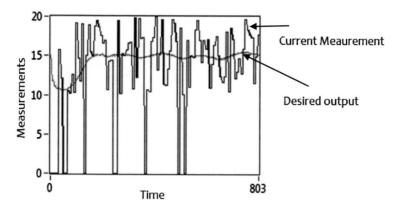

Fig. 10.14 LabVIEW Implementation of ANN block: It shows the output of the selected sensor node. Lesser the gap between the two *lines*, higher is the trust rating

---

ALGORITHM 10.1   NEURAL NETWORK ALGORITHM IN WSN

Input:
n: Number of sensor nodes
α_r: Learning rate = 0.5
Maintain history length
Sensor measurements $(x^{(1)}, x^{(2)}.....x^{(n)})$, where $x^{(i)} \in R^{(n)}$
Output: Trust ratings for the nodes.
1 **Data** acquisition
2 Select a random node A and store the value as x
2 **for** $i \leftarrow$ 1 *to* n except A **do**
4   Pass the current value of *i* as the input and the mean of the past information of *i* as the
            desired output to the ANN block
5 Store output of the ANN block as = $x_{predicted}$
6 Calculate mean of $x_{predicted}$
7 Calculate errors (Error) as x-$x_{predicted}$
8 Compare error with the threshold (τ) set as = 0.05
9   **if** *Error* $< \tau$ **then**
10   Node A has high trust rating
11   **Else**
12   Node A has low trust rating

(g) Fuzzy Logic

IF-THEN rules are used to solve a problem in fuzzy logic. The logical steps followed in the fuzzy logic approach are fuzzy sets and the criterion of selection is predefined along with the input variables such that the fuzzy rules give output for the given input data. The results are regularly updated with the feedbacks.

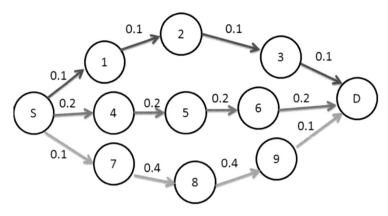

Fig. 10.15 Fuzzy Logic Calculation: 7, 8 and 9 are the most trustworthy nodes

Fuzzy logic security approach in WSN aims at finding the most trustworthy path using three principles namely (Kim and Seo 2008):

(1) *Fuzzy Matching*: calculates trust or mistrust degree.
(2) *Inference*: calculates rules and conclusion based on a matching degree.
(3) *Combination*: combines conclusion inferred by all fuzzy rules into the final one.

Initially, trust is calculated for each sensor node in the network followed by the summation of the trust values and combining them to select the best path as shown in Fig. 10.15. Figure has a source (sensor node) having three paths to reach to the destination (base station). The most reputable path has nodes numbered 7, 8, and 9 which have trust equal to 1 as against 0.4 and 0.8 of the other paths.

A TRIP mechanism was proposed which aimed at quickly and accurately distinguishing the fraudulent nodes spreading false information throughout the network (Mármol and Pérez 2012). RFSTrust based on fuzzy recommendation was proposed which comprises five types of fuzzy trust recommendation relationships according to the fuzzy relation theory and a mathematical description for MANETs. This approach not only provided a natural framework to deal with uncertainty but also the tolerance towards indistinct data inputs for the subjective tasks of trust evaluation, packet forwarding review and credibility adjustment (Luo et al. 2009).

The major problem in fuzzy logic approach is the memory overhead. Additionally, inefficiency occurs due to a lot of if-else rules.

(h) Swarm Intelligence

Swarm intelligence method also tries to achieve the most trustworthy path leading to the most reputable nodes in WSN (Mármol and Pérez 2011). Swarm intelligence approach uses ants as the inspiration for finding the most trustworthy nodes

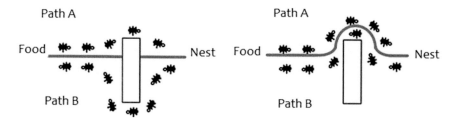

Fig. 10.16 Ants converging to shortest path (*Source* Perretto and Lopes 2005)

in the network. The main task of ants is to collectively search for food. They can adapt to the change in the environment, optimizing the path between the nest and the food source. This fact is due to stigmergy which involves positive feedback, given by the continuous deposit of chemical substance, known as pheromone. Classic example of searching for food via shortest path can be seen in Fig. 10.16.

Ants converge to the shortest path by selecting the path which has a higher deposition of pheromone. Pheromone is the physical substance secreted by the ants. As seen from Fig. 10.16, initially, half of the ants take one path and the other half take another path to reach to the food source. On their way, they deposit pheromone. The ants which first reach the food source are the first ones to come back again and give this information to other ants. Hence, the amount of deposition in the shorter path would be higher (see Eq. (10.20)) (Dorigo and Gambardella 1997).

$$\Delta \tau_{ij}^k = \begin{cases} \frac{Q}{L_k} & \text{if ant } k \text{ used edge } (i, j) \text{ in its tour} \\ 0 & \text{otherwise} \end{cases} \qquad (10.20)$$

where $\Delta \tau_{ij}$ is pheromone deposited from node i to node j and Q is a constant. The amount of pheromone deposition is inversely proportional to the length. Longer the path, pheromone deposition is less and shorter the path pheromone deposition is high. As time passes by, pheromone is updated using following equation) (Dorigo and Gambardella 1997):

$$\tau_{ij} \leftarrow \tau_{ij} + \sum_{k=1}^{m} \Delta \tau_{ij}^k \qquad (10.21)$$

On a similar note, pheromone is evaporated with the evaporation rate of ρ as depicted in the equation (Dorigo and Gambardella 1997):

$$\tau_{ij} \leftarrow (1 - \rho)\tau_{ij} \qquad (10.22)$$

By applying the above technique in WSN, computation of the most trustworthy path leads to the most reputable nodes in WSN. For implementation, 10 random active nodes were taken in a grid. Shortest path was computed having the most trustworthy nodes between the source and the destination as shown in Fig. 10.17

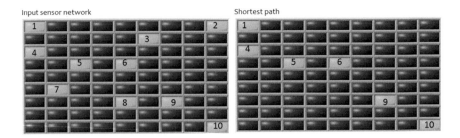

Fig. 10.17 LabVIEW implementation of swarm intelligence approach in WSN for finding the most reputable path

where 1 is the source node and 10 is the destination node and the path containing nodes 4, 5, 6 and 9 is the most trustworthy within the network.

Overall the method was efficient with 90 % detection rate of fraudulent nodes. However, the worst case space complexity of maintaining pheromone table is $O(n^2)$ for a fully meshed network of n number of nodes (Bhaskaran et al. 2011).

10.8.3   Comparative Analysis

Trust models are efficient in detecting fraudulent nodes in the network. Various metrics can be used for comparing the trust models as described below:

(1) *Indirect information requirement*: The first metric is whether indirect or second-hand information is required in computation of trust. Probabilistic method, game theory, fuzzy logic, and swarm intelligence do not require second-hand information. Rest of the methods requires second-hand information for computation of trust.

(2) *Space complexity*: Space complexity is defined as the amount of memory required in computation of the model. Ratings, weightings, probabilistic, fuzzy logic is less complex having a complexity $O(n)$, where n is the number of nodes. The space complexity increases in case of swarm intelligence and game theoretical approach. Neural network model has further high complexity $O(RT)$ *per cycle* where R is the total number of neurons and T is the maximum number of activation changes. Bayesian network approach was dependent on the number of components used which in turn generated complexity of $O(cn)$, where c is the number of components.

(3) *Time complexity*: Time complexity is defined as the amount of time required for the detection of fraudulent nodes in the network. It was lowest in case of Bayesian network model as bayes rule was used for detecting the fraudulent

nodes. It has the time complexity as $O(1)$. Probabilistic method also has less time complexity of $O(k)$ where k is the history length. Ratings, weightings has a time complexity of $O(n)$, where n is the number of nodes. Neural network has two time complexity associated with it (a) at training time and (b) at test time. At training time the complexity is of $O(f(\alpha_r))$ where α_r is the learning rate. Once the training is performed the time complexity of the model turns out to be $O(1)$. Game theory has a time complexity of $O(s_t^3)$ where s_t is the number of states in the game. This is because; Nash equilibrium is used for getting the best states. Nash equilibrium uses matrix row and column operations which is of the $O(s_t^3)$ (Wikipedia 2013). Fuzzy logic has a complexity of $O(n + e)$ where e is the number of edges in the network (Horowitz and Sahni 1978). Swarm intelligence complexity was dependent on the size of the network which in turn limits the application.

(4) *Efficiency*: Efficiency is defined as productivity of the system in detecting fraudulent nodes in the network. Neural network approach has output error less than 0.2 % which means the efficiency is 99 %. Swarm intelligence and probabilistic method also have high efficiency in detection of fraudulent nodes. It was lowest in case of ratings and weightings method.

(5) *Miscellaneous*: Besides the above metrics, each method has some problems that are independent from other models. While, ratings method was too simple and was applicable only for smaller networks; weightings method was inefficient for weight assignment. It worked under forwarding node notion which is not efficient in comparison to other models. Neural network despite of its high productivity in fraudulent node detection, suffered from high complexity which in turn is energy consuming and not recommended for resource constrained WSN nodes. Bayesian network approach was applicable only on flat architectures. Thus, the system was not scalable as more number of nodes could not be added in the network. Probabilistic method used only past information for the computation of trust. Fuzzy logic method suffered from memory overhead and inefficiency occurred due to a lot of if-else rules. Swarm intelligence suffered from colluding data node attack.

The comparative analysis of various trust models is illustrated in Table 10.2. The table shows among all the trust model, swarm intelligence which is a bio-inspired model, has the lowest complexity and highest efficiency. The model even deals with colluding data node attack where the number of fraudulent nodes is higher than the benevolent nodes in the network. Ratings and weightings method also had low complexity but the efficiency is very low.

The later sections in this chapter gives details on once the node is detected as fraudulent then what the network should do in respond to it. The method which is used for removing the fraudulent nodes in the network is a bio-inspired model and is a very effective model since it provides the solution depending upon the type of node.

Table 10.2 Comparative analysis of different trust and reputation models

Method	Indirect information	Space complexity	Time complexity	Efficiency (%)	Miscellaneous
Ratings	Yes	$O(n)$	$O(n)$	<25	Applicable for small networks
Weightings	Yes	$O(n)$	$O(n)$	25–30	Weight assignment
Neural network	Yes	O (RT)/cycle	$O(1)$	99	Highly complex, energy consuming
Bayesian network	Yes	$O(cn)$	$O(1)$	30–70	Applicable only on flat structure, not scalable
Probabilistic	No	$O(n)$	$O(k)$	80	Uses only past information
Game theory	No	$O(n^2)$	$O(s^3)$	30–60	High complexity
Fuzzy logic	No	$O(k^2)$	$O(n + e)$	45	Memory overhead and inefficiency due to lot of if-else rules
Swarm intelligence	No	$O(n)$	Depends on network size	90	Can capture colluding data node attack

Note n is the number of node, c is the number of components in Bayesian Network
R is the total number of neurons and T is the maximum number of activation changes
k is the history length
s is the number of states, e is the number of edges in the network

10.9   Bio-inspired Security Model for Wireless Sensor Networks

Biologically inspired security model for WSNs can be divided into two essential blocks namely machine learning model and immune model as shown in Fig. 10.18. Machine leaning model can be considered as brain performing learning and computation. Machine Learning model has three essential blocks namely clustering, support vector machine (SVM) and anomaly detection engine (ADE). It is used for the detection of fraudulent nodes. Machine learning model is followed by the immune model which is used to remove the fraudulent nodes from the system just like the defensive mechanism is used to remove foreign particles in human body.

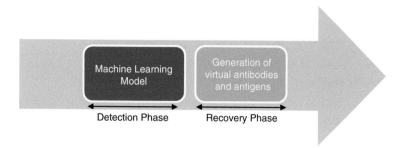

Fig. 10.18 Two phase model namely machine learning model (functioning as brain) and immune model

10.9.1   Machine Learning Model

Machine learning model is divided into two phases namely training phase and test phase as shown in Fig. 10.19: where training phase is used to train the data and formulate the machine to learn the trend, test phase is used to check the hypothesis made by the training phase. The output of the training phase is the set of all

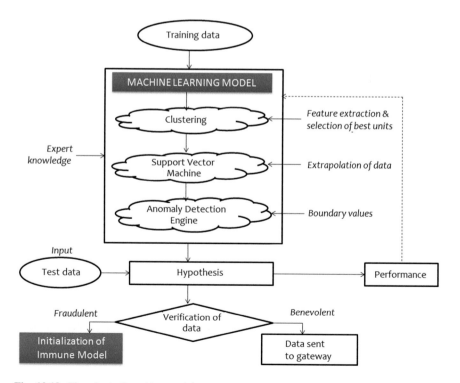

Fig. 10.19 Flowchart of machine model

fraudulent data measurements which is detected by the techniques of machine learning. Machine Model has three basic parts:

(1) *Clustering*
(2) *Support Vector Machine (SVM)*
(3) *Anomaly Detection Engine*

Clustering: It is an unsupervised machine learning method which works on the principle of finding a structure out of an unlabeled data set. The intent of unsupervised learning is to find a structure on unlabeled data set. K-means is a widely used technique under unsupervised learning. It is a type of algorithm which groups the data to produce clusters (Wagstaff 2001). The procedure follows a simple and effective way of creating clusters via two major tasks:

(1) *Cluster assignment*: Here random centroids (consider two centroids) are assigned and distance of each observation is measured with respect to the two centroids. Accordingly, assignment of each observation is made to the nearest centroid.
(2) *Centroid movement*: Calculate the new means to be the centroids of the observations in the new clusters, i.e., calculate the average of all the observations assigned to each centroid. Later assign the average position of these observations to the centroid.

Finally, repeat the above two steps till convergence. Algorithm 10.2 describes the k-means algorithm.

ALGORITHM 10.2   K-MEANS ALGORITHM

Input:
  k : *Number of clusters*
  s: Number of measurements
  Training set $(x(1), x(2)..... x(s))$, where $x(i) \in R(n)$
Output: k number of clusters generated out of random data measurements
 1 Randomly initialize k cluster centroids $\mu 1, \mu 2........\mu K$,
 2 **while** *till convergence* **do**
 3   **for** $i \leftarrow 1$ *to s* **do**
 4   c^i = index from 1 to k of cluster centroid closest to x^i
$$c^i = min_k \mid \mid x^{(i)} - \mu_k \mid \mid$$
 5   **for** $i \leftarrow 1$ *to s* **do**
 6   μ_k = average (mean) of measurements assigned to cluster k.

K-means method can be used to take the mean of the training data set to prepare a lower and an upper bound defining the two clusters. The two centroids which are assigned in k-means are the higher and the lower bound mean values of the data measurements. It groups the data to make clusters. Figure 10.20a shows the random measurements and Fig. 10.20b shows two clusters generated out of it using k-means.

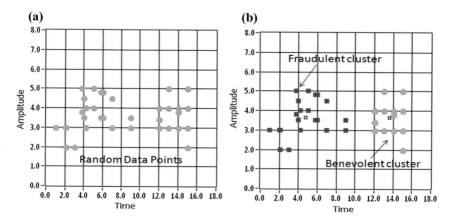

Fig. 10.20 K-means Algorithm for random data: **a** random data measurements are fed into k-means algorithm; **b** Two clusters generated out of k-means algorithm

Support Vector Machine: The data which was classified into clusters from k-means algorithm is given to SVM to create the decision boundary as shown in Fig. 10.21. Support vector machine (SVM) is a popular tool used to classify the data and create a decision boundary to distinguish between fraudulent and benevolent users/data. It is a type of classification and regression prediction tool. Over-fitting of data is avoided to maximize predictive accuracy by utilizing machine learning theory. The aim of SVM classifier is to determine a set of vectors called support vector to construct a hyper plane in the feature spaces (Tong and Koller 2002). Initially, the input vectors or features of the training data were

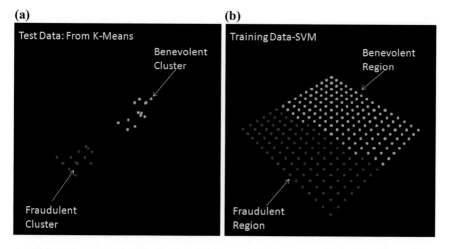

Fig. 10.21 3D view of support Vector machine: **a** The output of the k-means is fed into SVM; **b** Extrapolation of the measurements

mapped onto a very high-dimensional feature space. The mapped feature space, SVM then constructs a linear decision function which separates the two classes of events, namely benevolent and fraudulent. SVM searches for a decision function with maximum margins (Joseph et al. 2011).

Given the training datasets, $(x_i, y_i, i = 1, 2 \ldots n, y_i \in \{-1, +1\}, x_i \in R^d)$, a hyper plane is determined having a maximum margin (Suykens and Vandewalle 1999):

$$w_y.x = b \tag{10.23}$$

where w_y is a normal vector and the parameter b is offset. In order to find the optimal hyper plane, the following convex optimization problem is solved:

minimize
$$\left(\frac{||w_y||^2}{2} + \theta \sum_{i=1}^{n} \epsilon_i \right) \tag{10.24}$$
subject to
$$y_i(<w_y, x_i> +b) \geq 1 - \epsilon_i, \epsilon_i \geq 0, 1 \leq i \leq n$$

where $\sum_{i=1}^{n} \epsilon_i$ relax the constraints on the learning vectors, and θ is a constant that controls the tradeoff between number of misclassifications and the margin maximization. Equation 10.24 can be dealt using Lagrange classifier (Scholkopf and Smola 2001):

maximize
$$L(\alpha) = \sum_{i=1}^{n} \alpha_i - \frac{1}{2} \sum_{i=1}^{n} \sum_{j=`}^{n} \alpha_i \alpha_j y_i y_j K(x_j, x_i) \tag{10.25}$$
subject to
$$\sum_{j=1}^{n} y_i \alpha_i = 0 \text{ and } 0 \leq \alpha_i \leq c \text{ for all } 1 \leq i \leq n$$

where $K(x_j, x_i)$ is the kernel function and α_i is the Lagrange multiplier. According to the condition of Kuhn–Tucker (KKT), the x_is that correspond to $\alpha_i > 0$ is called support vectors (SVs). Once the solution to Eq. (10.25) is found, following equation is obtained:

$$w_y = \sum_{i=1}^{n} \alpha_i y_i x_i \tag{10.26}$$

Thus, the decision function can be written as:

$$f(x, a, b) = \{\pm 1\} + \text{sign} \left(\sum_{i=1}^{n} y_i \alpha_i y_i K(x, x_i + b) \right) \tag{10.27}$$

The kernel functions can be of various types, such as linear kernel, polynomial kernel, radial basis function kernel, sigmoid kernel, Gaussian kernel, convex optimization of kernels, normalization of kernels (Vapnik 2000; Muller 2001).

For the present study, linear kernel and polynomial kernel are used. They can be represented in following manner:

$$\text{Linear Kernel} \quad K(x_j, x_i) = <x_j, x_i>$$ (10.28)

$$\text{Polynomial Kernel} \quad K(x_j, x_i) = (s<x_j, x_i> + c)^d$$ (10.29)

where s, c, and d are kernel specific parameters.

Anomaly Detection Engine: SVM creates the decision boundary; however the data which lies on the boundary needs to be further evaluated for better accuracy and precision. ADE helps in making more accurate decisions. The algorithm works in the following manner.

- After SVM implementation, the Gaussian distribution of benevolent features (x_j, y_j) is calculated that might be indicative of anomalous examples for clear cut demarcation between the fraudulent and the benevolent data.
- The Gaussian distribution is calculated by computing mean and standard deviation of the benevolent features.

$$\mu_j = \frac{1}{S} \sum_{i=1}^{S} x_j$$ (10.30)

$$\sigma_j^2 = \frac{1}{S} \sum_{i=1}^{S} (x_j - \mu_j)^2$$ (10.31)

where μ and σ are the mean and the standard deviation of the benevolent measurements.

- Compute the probability distribution $p(x)$ with the help of following equation (Ng 2012a, b):

$$p(x) = \prod_{j=1}^{n} p\left(x_j, \mu_j, \sigma_j^2\right) = \prod_{j=1}^{n} \exp \frac{-(x_j - \mu_j)^2}{2\sigma_j^2}$$ (10.32)

- The combined probability distribution function of x and y is calculated using the following equation:

$$p(x) = p\left(x, \mu_x, \sigma_x^2\right) p\left(y, \mu_y, \sigma_y^2\right)$$ (10.33)

- Threshold ϵ is maintained and if the probability of the measurement is $<\epsilon$, it is considered as anomaly and vice versa.

The probability distribution function of x and y and combined probability is as shown in Fig. 10.22. The outcome of the machine learning model is the set of fraudulent data measurements. The algorithms for the training and the test phase of machine learning model are described in Algorithms 10.3 and 10.4 respectively.

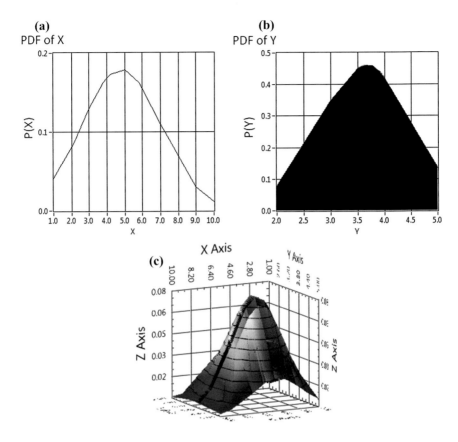

Fig. 10.22 Probability distribution function of x, y and combined probability are shown in (**a**), (**b**) and (**c**) respectively

---

ALGORITHM 10.3   TRAINING PHASE OF MACHINE LEARNING MODEL

Input:
    m: Number of measurements
    Training set $((x^{(1)}, y^{(1)}), (x^{(2)}, y^{(2)})..... (x^{(m)}, y^{(m)})$ where $(x^{(i)}, y^{(i)}) \in R^{(n)}$. Here, $x^{(i)}$ denotes time and $y^{(i)}$ denotes measurements.
Output: Set of all fraudulent data measurements detected by the techniques of machine learning
1 Data acquisition
2 Clustering
3 Support vector machine
4 Anomaly detection engine
5 **for** *each measurement in benevolent set* **do**
6   **if** probability < threshold (0.1) **then**
10   *Anomaly*
8   **Else**
9   *not an anomaly*

ALGORITHM 10.4	TEST PHASE OF MACHINE LEARNING MODEL

Input:
Run time data $((x^{(1)}, y^{(1)}), (x^{(2)}, y^{(2)})..... (x^{(m)}, y^{(m)}))$ where $(x^{(i)}, y^{(i)}) \in R^{(n)}$.
g: Flag of antigen
Output: Detection of fraudulent node
1 Data acquired at run time
2 **for** *each new measurement* **do**
3   **if** *acquired measurement is in benevolent set* **then**
4   *Set g = 1, (Initialization of immune model)*
5   **Else**
6   *send measurement to the training set for the update*

10.9.2 Immune Model

Immune model guides a way to deal with fraudulent nodes which are detected as fraudulent nodes as an outcome of the machine learning model. The flowchart for the same is shown in Fig. 10.23.

A mathematical model was devised to study the overall antigen-antibody concept (Dibrov et al. 1977a, b; Fowler 1981). Dibrov Model consists of three coupled equations for the antibody quantity a, the antigen quantity g, and the small B-cell population x. If the secondary and subsequent responses to an antigen injection are

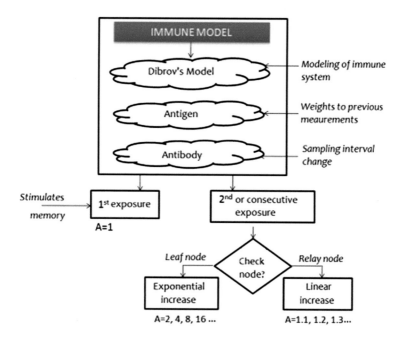

Fig. 10.23 Flowchart of immune model

similar, it is reasonable to suppose that the small B-cell population is relatively unaffected by the later injections, so x is put as a constant. The second-order set of equations describing antigen-antibody interactions:

$$\frac{dg}{dt} = Kg - Qag \qquad (10.34)$$

$$\frac{da}{dt} = AH(t - T) - Rag - Ea \qquad (10.35)$$

where K, Q, A, R, E are rate constants. K is the overall growth rate of antigen. $H(t)$ in Eq. (10.35) is the Heaviside step function whose value is zero for a negative argument and one for a positive argument.

$$H(T) = \begin{cases} 0 & t < 0 \\ 1 & t \geq 0 \end{cases} \qquad (10.36)$$

The product 'ag' is the complex formed as antibody-antigen complex. As the complex is formed, it results in a net loss of the antibody and antigen. The simplest assumption is that of the law of mass action, valid when the densities are below a certain saturation level, i.e., the losses are proportional to the product of the antibody and antigen densities. The rate constants Q and R are necessarily not the same. The rate of antibody production at time t is supposed to be proportional to the rate of small B-cell stimulation at time $t - T$: i.e., there is a delay T between the stimulation of a small B-cell and the subsequent production of plasma cells from it.

In the Immune model, Dibrov Model was taken as the foundation for the removal of fraudulent nodes. Simulations were carried out using the Runge–Kutta (variable) method for solving the Dibrov differential Eqs. (10.34) and (10.35) and the following results were obtained as shown in Fig. 10.24. It shows the graph of

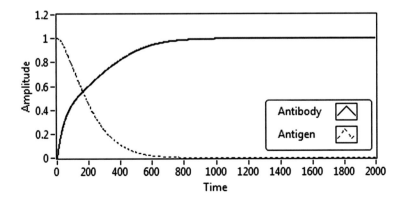

Fig. 10.24 Rate of change of antigen and antibody

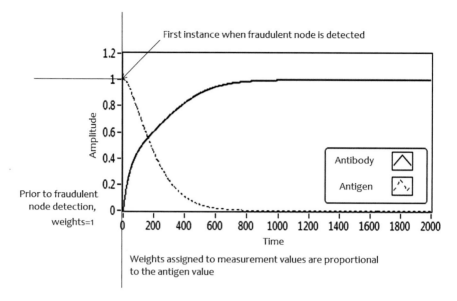

Fig. 10.25 Relationship between weights and antigens

the rate of change of antigen and antibody as a function of time, for values of $K = 0.01$, $Q = 1$, $A = 1$, $R = 1$, $E = 1$ with initial conditions $a_0 = 0$ and $g_0 = 1$. This shows that initially, the antigen count is high which decreases when antibodies are produced in the body. In other words, when an antigen is detected, the B-cells start producing antibodies to form an antigen-antibody complex. The count of antigen decreases linearly with the increase in antibodies.

Immune model in WSN focuses on giving correct measurements to the host controller even if the node is detected as fraudulent by the machine learning model. This is done so as to increase the lifetime of the node so that even if the node gets corrupted, it gives correct measurements. For this, weights are assigned to the measurement values of the fraudulent nodes. The weights assigned are proportional to the antigen values taken from the differential equations. Prior to the fraudulent node detection, the weight assigned is equal to 1. After the fraudulent node is detected, the expected measurement value relies upon the previous measurements and the weights would then be decreased proportionally to the antigen values taken from the differential equation as shown in Fig. 10.25.

The measurements are calculated as per the following equation:

$$T_{new} = \frac{\sum_{i=1}^{k} W_i \times T_{prev}}{k} \tag{10.37}$$

where T_{new} is the new measurement obtained by applying weights, T_{prev} are the previous measurements, k is the history length fixed as 10, W_i are the weights which are kept proportional to the antigen values as per the following equation:

$$W_i = \theta \times g_i \qquad (10.38)$$

where θ is a constant.

There are several ways to react to the fraudulent node. In order to minimize the effect of fraudulent nodes either the sampling interval can be increased or decreased. In WSN the measurements are acquired at a particular rate which is termed as a sampling interval. Prior to the fraudulent node detection the sampling interval is kept as constant. After the detection of fraudulent sensor node, it is required to reduce the effect of this node. However, turning off the fraudulent node immediately after detection is not a feasible solution since it would affect the stability of the system. So a better solution would be to either slowly decrease the sampling interval to zero or to increase the sampling interval depending upon the application. Increasing the sampling interval will not only help the host controller to check for on-off attacks but also help in increasing the lifetime of the fraudulent sensor node. The type of an attack, where the sensor node changes its behavior alternatively is known as an on-off attack (Lopez et al. 2010). Increasing the sampling interval would allow the good measurements to last for a longer period since the rate at which the samples are collected gets increased.

The sampling interval is increased or decreased by taking into account the antibodies' value from the differential equation. The rate at which the sampling interval is increased is:

$$s_{\text{after\_attack}} = 2^{\left\lfloor \dfrac{\frac{a_i}{a_{\max} - a_{\min}}}{k} \right\rfloor + 1} \times s_{\text{prior\_to\_attack}} \qquad (10.39)$$

$s_{\text{after\_attack}}$ is the sampling interval after the fraudulent node is detected and $s_{\text{prior\_to\_attack}}$ is the sampling interval prior to the detection of the fraudulent node, a is the antibody value where $a_{\max}$ and $a_{\min}$ is fixed to 1 and 0 respectively. k is the number of steps desired to end the influence of fraudulent node (in this case it is fixed to 10).

The rate at which the sampling interval is decreased is:

$$s_{\text{after\_attack}} = \dfrac{s_{\text{prior\_to\_attack}}}{2^{\left\lfloor \dfrac{\frac{a_i}{a_{\max} - a_{\min}}}{k} \right\rfloor + 1}} \qquad (10.40)$$

Furthermore, the feature of the primary and the secondary response can be adapted in shutting down the fraudulent sensor nodes. In WSN, the shutting down process of the fraudulent node can depend upon the number of backup nodes:

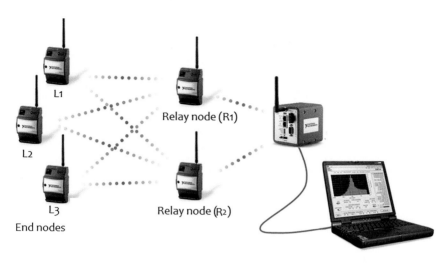

Fig. 10.26 Leaf versus relay node for shutting process

(1) *Leaf Node*: Do not hamper the network stability. The response time to shut the leaf node should be high.
(2) *Relay Node*: Can hamper the network stability. It should be removed slowly.

Consider Fig. 10.25 which shows node can be either a leaf node or a relay node. In Fig. 10.26, L1, L2, and L3 are the leaf nodes and R1 and R2 are the relay nodes. Since the fraudulent node can either be a leaf node or a relay node, shutting off those nodes should be done proportionally. If the node is a leaf node it can be removed quickly, however, if it is a relay node, the response time to shut off that node should be high since many leaf nodes would be dependent on it. This is done so as to improve the stability of the system. For instance, R2 becomes fraudulent at some moment then L3 node should route its packet to some other node other than R7. Therefore, shutting down process would depend on the number of backup nodes.

In order to adapt this feature for shutting down process, A constant used in Eq. 10.35 (rate of change of antibody) is increased either linearly or exponentially depending on the type of the node. The relay nodes which are detected as fraudulent should be removed slowly from the network. Linear response can be an effective solution in this regard. Linearity can be depicted by the following equation:

$$A(t) = \begin{cases} 1 & n = 1 \\ 0.1 + A(t-1) & n \geq 2 \end{cases} \qquad (10.41)$$

where $A(t)$ is the increase at time instant t, n is the number of exposure.

For shutting down the leaf nodes, exponential increase in (A) is an effective solution. Exponential increase of antibody (a) is mathematically modeled by the following equation:

$$A(t) = A(t-1)[1+r]^n \tag{10.42}$$

where r is the growth rate. Keeping $r = 1$, (A) is increased exponentially by following equation:

$$A(t) = \begin{cases} 1 & n = 1 \\ 2^n A(t_0) & n \geq 2 \end{cases} \tag{10.43}$$

where $A(t_0)$ is the initial antibody count, i.e., equal to one. Algorithm 10.5 illustrates the overall algorithm for immune model.

ALGORITHM 10.5   IMMUNE MODEL

Input:
  $x^{(k)}$, $y^{(k)}$ as the measurement of a fraudulent sensor node
    g: *Flag for initialization.*
Output: Turning off the fraudulent node slowly by ignoring its reading.
1 Set $g(t) = 1$ for that particular sensor node for others this parameter is zero
2 Solve Dibrov's differential equation.
3 Update the measurements of fraudulent node by assigning weights to its previous
   measurements using antigen values
4   **if** *first exposure* **then**
5   Set $A = 1$
6   **Else** *second exposure*
7   **if** *leaf node* **then**
8   *linearly increase A*
9   **Else** *relay node*
10   *exponentially increase A*
11 Change the sampling interval using values received from antibody production.

10.9.3   Experiments and Results

Simulations were carried out in LabVIEW, on a network size of 10 sensor nodes out of which, one sensor node was introduced with high noise (See Fig. 10.27). Noise faults are introduced where the temperature readings $T_{x,y}$ were replaced by $T_{x,y} + x_g$, where x_g was a Gaussian random variable, whose distribution was $N(0, 2)$. Controlled environment setup is taken into account where the temperature is consistent around 20°–30 °C as shown in Fig. 10.28. Any value outside this temperature range is considered as a fraudulent behavior.

(a) Machine Learning Model

The sensor nodes were trained for 24 h, where machine learning tools like k-means, SVM, and ADE were used. The k-means was used to create two clusters namely fraudulent cluster and benevolent cluster as shown in Fig. 10.28b on account of training data. The data which was classified into clusters was given to SVM which

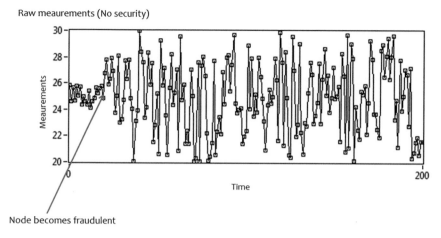

Fig. 10.27 Fraudulent sensor node measurements (Noise Faults)

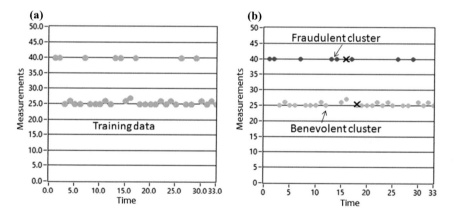

Fig. 10.28 K-means Clustering: **a** Training set having temperature measurements; **b** Shows the learned data having two clusters generated out of k-mean algorithm having centroid assigned as mean value

created the decision boundary. Linear kernel and polynomial kernel are used for the simulations. However, polynomial kernel is preferred in this case where measurements till 36 °C were treated as benevolent measurements. This is because if the same type of data comes continuously it will be considered as a benevolent data as seen from Fig. 10.29. The data which lies on the boundary needs to be further evaluated for better accuracy and precision as shown in Fig. 10.29. ADE is used for the boundary values between the two regions as shown in Fig. 10.30.

(b) Immune Model

Immune model is instantiated when the sensor node sends a fraudulent measurement which lies in the data set of fraudulent measurements as classified by the

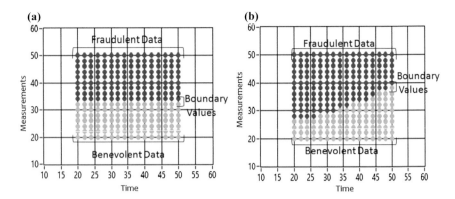

Fig. 10.29 SVM Linear and polynomial output: **a** and **b** shows the extrapolation of data measurements using SVM linear and polynomial kernel respectively. Polynomial kernel is used for further processing because if the same type of acceptable data comes continuously it will be considered as a benevolent data as seen from the curve

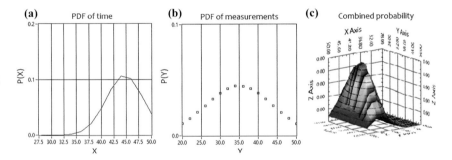

Fig. 10.30 Probability distribution function of the benevolent measurements of SVM polynomial kernel: **a** PDF of x axis of the polynomial kernel i.e. the time; **b** PDF of y axis i.e. the temperature measurements. It explains the probability of having 25 °C is 0.03; **c** Combined probability of the two

machine learning model. In an immune model, four different scenarios were considered for the purpose of comparison (See Fig. 10.31):

(1) *Non-weighted averaging*: Simple averaging is performed, not assigning weights to the previous measurements. Here, the measurements would be varied according to the noise in the fraudulent node. Hence, the variance from the true measurement gets varied. Lifetime of the fraudulent node is infinity here, since there is no change in the sampling interval for capturing the future measurements.

(2) *Weighted averaging*: Weights are assigned to the previous measurements for the computation of new measurement readings. Here, the measurements would decrease since it is dependent on the weights given to the previous measurements as per Eqs. (10.37) and (10.38). Weights are made proportional to the antigen values taken from the differential Eqs. (10.34) and (10.35).

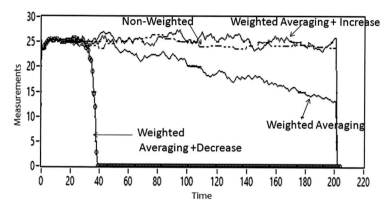

Fig. 10.31 Measurements on Applying the Concept of Antigen and Antibody: Clearly it is seen that measurements would be good for some time even if the node becomes fraudulent

(3) *Weighted averaging and increase the sampling interval*: Besides weighted averaging, sampling interval is increased using Eq. (10.39). In this case, good measurements persist their state for longer duration making the lifetime of fraudulent node longer. The sampling interval is increased by taking into account the antibodies' value from differential Eqs. (10.34) and (10.35).

(4) *Weighted averaging and decrease the sampling interval*: Weighted averaging is performed and sampling interval is decreased using Eq. (10.40). Here, variance from true measurements would be high which leads to a shorter lifetime of the fraudulent node. Also, when the sampling interval is changed, sampling interval of other channels would also get affected.

Figure 10.30 shows the results over the four scenarios. It can be seen that soon after 20 s, when the sensor node starts giving fraudulent measurements (Fig. 10.26), host controller was able to get good measurements for a long time with the help of immune model. In case of weighted averaging and increase in sampling interval the lifetime of getting good measurements was high. However, with the increase in sampling interval, the measurements lasted for 200 s. Furthermore, in case of weighted averaging and decrease in sampling interval, host controller was getting good measurements for 20 more seconds and then the node was getting turned off.

Table 10.3 gives the comparison to the four scenarios discussed above. As seen from Table 10.3, comparison is made on three metrics, i.e., lifetime of fraudulent node, variance from true measurement and impact on other channels. Lifetime of fraudulent node should be less and in case of weighted averaging and decrease in sampling interval lifetime comes out to be the least. Variance from true measure-ment is less in case of weighted averaging and increase in sampling interval. The impact on other channels only happens in the last two cases. As seen from the results it is recommended to use weighted averaging. The sampling interval can

Table 10.3 Results comparing four different scenarios

Metric	Non-weighted averaging	Weighted averaging	Weighted averaging + increase sampling interval	Weighted averaging + decrease sampling interval
Fraudulent node lifetime (response time)	Infinity	Medium	Medium	Low
Variance from true measurement	Varied, depends on type of noise	High	Low	High
Impact on other channels in infected node	None	None	High	High

either be increased or decreased depending upon the application. Evidently, the fraudulent node should be removed after its detection, for that reason weighted averaging and decrease in sampling interval is recommended. However, fraudulent node can be monitored for longer time to check for an on-off attack, where weighted averaging and increase in sampling interval would be preferred.

Weighted averaging and decrease in sampling interval (WADS) can be compared with the filter response. There are three bands in a filter namely passband, transition band, and stop band. Passband response is the filter's effect on frequency components that are passed through. Frequencies in the middle which are not removed completely from the output signal, may receive some attenuation are represented by transition band. Frequencies in the stop band are removed completely from the output signal. Similar type of analogy can be established for the sensor nodes by assigning a pass band, a transition band, and a stop band. These three phases can be compared very beautifully with the WADS approach as shown in Fig. 10.32. There would be a passband where measurements would be taken from the fraudulent sensor node followed by the transition band giving the network some time to stabilize. Stop band is the phase where the measurements would be ignored totally.

Furthermore, primary and secondary responses of immune system can be adapted in WSN depending on the type of node. It was observed that while increasing the value of A as a linear function Eq. (10.41); the response time was low as shown in Fig. 10.33a where time is in seconds. However, as A was increased exponentially Eq. (10.43), the response time was fast as shown in Fig. 10.33b. The figure explains that first exposure is same in both cases but it changes its behavior on second exposure depending on the type of the node.

The present section provided details on how biologically inspired techniques can prove as a benefactor in security for WSN. The following section intelligent water drops would suggest what recovery path mechanism should be applied once the nodes are removed from the system for better reliability. As the fraudulent nodes

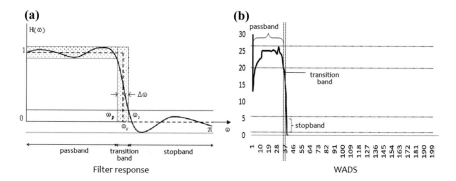

Fig. 10.32 Filter Response and WADS Technique: **a** Filter response; **b** Weighted averaging and decreasing sampling interval can be compared beautifully with the filter response; there would be a passband, a transition band and a stop band where measurements would be neglected fully

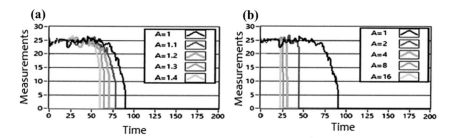

Fig. 10.33 Secondary response: **a** Linear increase of A on relay node. On second exposure the time required to turn off the fraudulent relay node is around 77 s versus; **b** Exponential increase of A on leaf node. On second exposure the time required to turn off the leaf node is only 48 s

are removed from the network, the connecting nodes should pass its data to the other nodes. For this, another bio-inspired technique called as intelligent water drops has been proposed which determines various alternative paths that can be applied in wireless sensor networks (Fig. 10.34).

Fig. 10.34 Topology of the sensor network

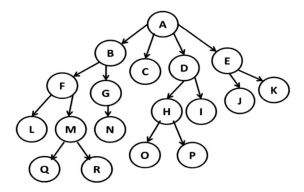

10.10   Intelligent Water Drops

IWD algorithm has been studied extensively which is nothing but an idealized approach of studying natural water drops in rivers (Chen et al. 2009). Natural water drops have two properties that are soil and velocity. Every water drop flows with some velocity and carries an amount of soil with itself. The velocity of the water drop helps in removing some soil from the riverbeds, and this soil is added to the soil of the water drop. During the transition, the water drops gains speed; due to which a water drop with more velocity can erode more soil. One more interesting feature of the water drop is that it chooses an easy path when there are many paths available at the same time. Here, easy path means the path with lesser soil. In this way constructively, the water drop chooses its optimal path. Using these remarkable properties IWDs are created. These drops possess two important properties of natural water drops that are:

1. The soil it carries, denoted by soil (IWD)
2. The velocity that it possesses, denoted by velocity (IWD).

   Both these values can change when an IWD flows from one location to another in its environment (given problem statement). The IWDs' velocity is nonlinearly proportional to the inverse of the soil on the path between the two locations. The IWDs' soil is increased by removing some soil from the path, and the amount of soil added is nonlinearly proportional to the inverse of the time taken to move between the locations. This time is proportional to the velocity and inversely proportional to the distance between the two locations. In addition, the path choosing property is implemented by calculating an uniform random distribution function, such that the probability of choosing a path is inversely proportional to the soil of the available paths. This section briefs about IWDs, which is used to find optimized path of WSN to find the most trustworthy nodes in the network. The following section explains how the properties of water drops are applied to the desired problem and the process of achieving the reputable paths.

10.10.1   IWD in Wireless Sensor Networks

Topology of the network is known at the start. Initially, the important factors affecting the working of the algorithm are explained followed by a stepwise explanation of the algorithm.

1. Fitness Function

A fitness function has been designed for the calculation of soil on all the links of the graph (Chen et al. 2009). The basic idea behind using the fitness function is to give importance to more deep and complex paths. A well-constructed fitness function may increase the chances of finding a solution and reaching higher coverage, which is always desirable. For the proposed algorithm, the fitness function can be calculated as:

$$\text{soil} (i, j) = a * \text{subgraph} (j) + b * \text{condition} (j) \tag{10.44}$$

where soil (i, j) is the soil on the edge between the two nodes 'i' and 'j'.

Condition (j) of a node 'j' can be either 0 or 1, depending on whether a particular node has a decision element or not. If it has a decision statement then it is assigned a value of 1 else 0. Subgraph (j) is the subgraph value, i.e., number of nodes below that particular node (j) in a given graph.

The IWD starts from the initial node of the graph and after that the probability of each path is calculated using the above mentioned fitness function. The path with the highest probability is considered as the desired path for the water drop. The configuration parameter 'a' and 'b' is assigned a constant value of 2 and 1, respectively. The reason behind assigning these values is that, more weight should be assigned to nodes with a larger subgraph value rather than nodes with smaller subgraph value. Subgraph value has been given greater importance in the designing of the fitness function, because by traversing the part of the code with greater depth the probability of achieving complete code coverage is high. This fitness function was so chosen, because during code coverage there are chances of errors in either decision nodes or more dense and complex paths. In order to assign weight to links based on their complexity, this fitness function is chosen.

2. List of Visited Paths

A list of visited paths known as 'V_c' is maintained to record the paths traversed by the IWD. In order to keep track of the nodes visited, a check is performed on the node, and if all its children are visited then it is deleted. This operation is performed recursively. This is done in order to avoid the IWD from traversing the path which has been already traversed.

3. Constant Values

Various configuration parameters (a_s, b_s, c_s), (a_v, b_v, c_v), ρ are used in the calculation and updating of soil and velocity, respectively during implementation of the IWD algorithm (Shah-Hosseini 2007, 2009).

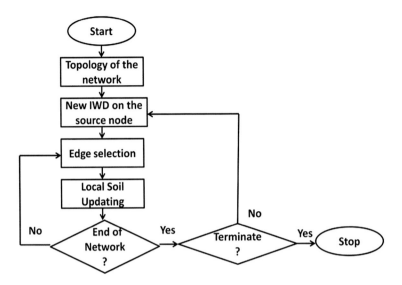

Fig. 10.35 Flowchart for the IWD algorithm

10.10.2 Algorithm

Parameters which are used are:

- Probability (i, j): probability of the IWD to move from node i to node j.
- Time (i, j): time taken by the IWD to move from node i to node j.
- ΔSoil (i, j, iwd): soil eroded by the IWD on the link between node i and node j.
- Vel (i, j, iwd): velocity of the IWD when it moves from node i to node j.

Step wise implementation of the algorithm is described in Algorithm 10.6 and Fig. 10.35 (Aggarwal et al. 2012).

Algorithm 10.6 is described with the help of example as shown in Fig. 10.34. In this topology, A can be considered as the source node from where data is transmitted. Sub-graph of all the nodes is calculated first. Subgraph value of a node is the number of nodes below that particular node. Node A has all the nodes, i.e., B, C, D, and so on connected to it. This is done using the depth first traversal algorithm (Cormen 2009). Hence the value of the sub-graph of all nodes is:

$$A = 17, B = 7, D = 4, E = 2, F = 4, G = 1, H = 2, M = 2$$
$$C, I, J, K, L, N, O, P, Q \text{ and } R \text{ have a sub graph value of } 0.$$

In this way, all the possible paths are determined with their corresponding weights. The path with highest weight should be the most trustworthy path. Next section shows the demonstration of the above mentioned algorithm on an illustrated problem.

The complete flowchart of the algorithm is:

STEP 1: INITIALIZATION OF PARAMETERS

a.   Velocity updating parameters are a_v, b_v, c_v. set as: $a_v = 1$, $b_v = 1$, $c_v = 1$.

b.   Soil updating parameters are a_s, b_s, c_s set as: $a_s = 1$, $b_s = 1$, $c_s = 1$.

c.   The local soil updating parameter is ρ. Here ρ = 0.1 is chosen for better experimental results. Value of ρ can range from 0 to 1, depending upon the amount of soil on the link (nodes).

d.   The initial soil on each edge of the graph is computed using the fitness function and the configuration parameters

e.   The initial velocity of each IWD is set to 100.

f.   Every IWD has a visited path list Vc (IWD) which is initially empty i.e. Vc (IWD) = {}.

STEP 2: UPDATION OF PARAMETERS VALUES

Repeat below steps until topology does not gets empty:

2.1. Calculation of probability:

For the IWD receding at node i, the next node j (which is not in visited node list Vc (IWD)) is chosen on the basis of probability $p_i^{IWD}(j)$. The probability for IWD to move from the current node i to node j is calculated using:

Probability (i, j) = Soil (i, j) / Σ k ≠vc Soil (i, k)

Where: Soil (i,j) is the soil between the two nodes, Σ k ≠vc Soil (i,k) denotes the summation of the soils of all the paths which can be traversed from the current node i. In the original algorithm, the IWD moves to the location where the soil is less, but in our case we traverse the IWD to the path with more soil.

2.3 Calculation of time taken:

After calculating the probability, IWD chooses the next node which it has to move on. After selecting the respective node, the time taken to move from node i to node j is calculated as:

Time (i.j) = (subgraph (i)–subgraph (j)) / vel (iwd)

Where subgraph(i) – sugraph(j) correspond to the distance between the nodes and vel(iwd) is the original velocity of the IWD.

2.4 Calculation of Soil of the IWD:

Soil carried by IWD is computed using 2.3:

ΔSoil (i, j, iwd) = a_s / (b_s+c_s*time (i, j))

Where: a_s, b_s, c_s are the positive parameters as specified earlier.

2.5 Calculation of Velocity of the IWD:

The velocity of the IWD after it has moved from node i to j is calculated as:

vel(i,j,iwd) = vel(iwd) +(a_v +(b_v+c_v soil(i,j)))

Where: soil (i,j) is the soil on the path before the IWD traversed the required path. a_v ,b_v, c_v are the positive parameters as specified earlier and vel(IWD) is the previous velocity of the IWD.

2.6 Soil updation of the link: Since, the IWD carries some amount of soil with it; the soil on the link is reduced. Therefore, the soil on the link is updated as:

Soil (i,j)= (1-ρ) soil (i,j) – ρ * Δsoil (i,j,iwd),

Where: ρ is a positive parameter, Δsoil (i, j, iwd) is the soil carried by the IWD while moving from the node i and j.

2.7 Current node is updated and the same procedure from step 2.2 is repeated until it does not reach leaf node, or one complete path is not reached. After it reaches the leaf node then that node is deleted, and its parent is also checked. If all its leaves have been deleted, then it is also deleted.

2.8 After completing one possible path by the water drop, the visited path list is updated and the IWD again starts with a new iteration where all parameters are initialized again from step 1 onwards. If the topology becomes empty, that means all the paths have been traversed, and we have to exit the loop else all the steps from 2.1 are executed.

Next, the decision condition for the node is checked. *A, B, D, E, F, G, H,* and *M* are decision nodes having value 1 and rest of them with a value of 0. The various iterations of the IWD are shown below.

Iteration 1: The IWD starts from initial node *A* which can be considered as the base station as shown in Fig. 10.36.

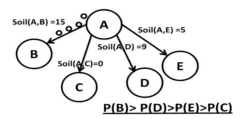

Fig. 10.36 Path selection by IWD from *A*. *Note* Visited path (Vc) = A − >B

As the water-drop, has four ways to travel, i.e., by means of *B, C, D,* and *E*. Along these lines, with a specific end goal to figure out which hub to be navigated next, the likelihood of two hubs is registered utilizing the particular connection soil and on the premise of the following hub the first path is chosen.

The calculation of the various parameters is

$$\text{Soil}\,(A,\,B) = 2*7+1*1 = 15$$
$$\text{Soil}\,(A,\,C) = 2*0+1*0 = 1(\text{as sub} - \text{graph}(j)\&\&\text{condition}(j) == 0)$$
$$\text{Soil}\,(A,\,D) = 2*4+1*1 = 9$$
$$\text{Soil}\,(A,\,E) = 2*2+1*1 = 5$$
$$P\,(A,\,B) = \text{Soil}\,(A,\,B)/\{\text{Soil}\,(A,\,B) + \text{Soil}\,(A,\,B) + \text{Soil}\,(A,\,B) + \text{Soil}\,(A,\,B)\}$$
$$= 15/30$$
$$P\,(A,\,C) = 1/30$$
$$P\,(A,\,D) = 9/30$$
$$P\,(A,\,E) = 5/30$$

The node with a higher probability is chosen first. In this case probability of node B is higher than that of node C, D and E so node B is selected. The probability is calculated using the Step 2.1 of the algorithm mentioned above. The time taken by the IWD to move from *A–B* is calculated using Step 2.2 of the algorithm mentioned above:

$$\text{Time}\,(A,\,B) = \max\{(17 - 7),\,1\}/\text{Velocity}\,(\text{Viwd})$$
$$= 10/100 = 0.1\,\text{s}$$

The soil carried by the IWD and its velocity is computed which is then subtracted from the soil of the link *A–B* to compute the remaining soil on the link using steps 2.3 and 2.4 and 2.5:

$$\Delta \, \text{Soil} \, (A, \, B) = 1/(1+1*0.1) = 0.9$$
$$\Delta \, \text{Velocity} \, (A, \, B) = 100+1/1+1*15 = 100.0625$$
$$\text{Updated Soil} \, (A, \, B) = 0.9*15{-}0.1*0.9 = 13.41$$
$$\text{Soil (iwd)} = 0.9$$

So the path will be: A–B.

Now, after reaching node B the water-drop has two available nodes for selecting the next node, which is shown in Fig. 10.37.

Again, the probability of all the available nodes is calculated and on the basis of that the next node is chosen.

$$\text{Soil} \, (B, \, F) = 2*4+1*1 = 9$$
$$\text{Soil} \, (B, \, G) = 2*1+1*1 = 3$$
$$P \, (B, \, F) = \text{Soil} \, (B, \, F)/\{\text{Soil} \, (B, \, F)+\text{Soil} \, (B, \, G)\}$$
$$= 9/12 = 0.75$$
$$P \, (B, \, G) = 0.25$$

As the probability of node F is greater so the water drop will flow to node F.

$$\text{Time} \, (B, \, F) = \{7-4\}/100.0625 = 0.0299$$
$$\Delta \, \text{Soil} \, (B, \, F) = 1/(1+1*0.0299) = 0.9709$$
$$\Delta \, \text{Velocity} \, (B, \, F) = 100.0625+1/1+1*9 = 100.1625$$
$$\text{Updated Soil} \, (B, \, F) = 0.9*0{-}0.1*0.970 = 8.003$$
$$\text{Soil (iwd)} = 0.9+0.970 = 1.87$$

So the path will be: A–B–F.

Similarly, after reaching node F the water-drop has to decide the next node to be traversed as in Fig. 10.38.

Fig. 10.37 Path selection by IWD from B

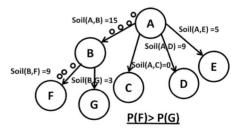

Fig. 10.38 Path selection by
IWD from *F* and *M*

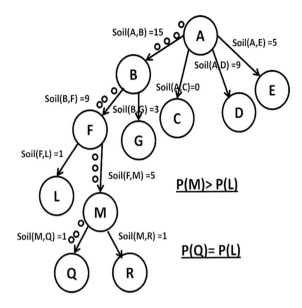

The calculations for the various parameters are shown below:

$$\text{Soil}\,(F,\,L) = 2*0+1*0 = 1$$
$$\text{Soil}\,(F,\,L) = 2*0+1*0 = 1$$
$$\text{Soil}\,(F,\,M) = 2*2+1*1 = 5$$
$$P\,(F,\,L) = \text{Soil}\,(F,\,L)/\{\text{Soil}\,(F,\,L)+\text{Soil}\,(F,\,M)\}$$
$$= 1/6 = 0.167$$
$$P\,(F,\,M) = 5/6 = 0.83$$

As the probability of node *M* is greater, the water drop will flow to node *M*

$$\text{Time}\,(F,\,M) = \{4-2\}/100.1625 = 0.01996$$
$$\Delta\,\text{Soil}\,(F,\,M) = 1/(1+1*0.01996) = 0.98$$
$$\Delta\,\text{Velocity}\,(F,\,M) = 100.1625+1/1+1*5 = 100.329$$
$$\text{Updated Soil}\,(F,\,M) = 0.9*5-0.1*0.98 = 4.402$$
$$\text{Soil}\,(\text{iwd}) = 1.87+0.98 = 2.85$$

So the path will be: A–B–F–M.

$$\text{Soil}\,(M,\,Q) = 2*0+1*0 = 1$$
$$\text{Soil}\,(M,\,R) = 2*0+1*0 = 1$$
$$P\,(M,\,Q) = P\,(M,\,R) = 1/2 = 0.5$$

Now, as the probability of two available nodes is same. A random selection is to be done among these three nodes. So, node Q is randomly chosen as the next node.

$$\text{Time } (M, Q) = \{2 - 0\}/100.329 = 0.01993$$
$$\Delta \text{ Soil } (M, Q) = 1/(1 + 1 * 0.0199) = 0.98$$
$$\Delta \text{ Velocity } (M, Q) = 100.329 + 1/1 + 1 * 1 = 100.829$$
$$\text{Updated Soil } (M, Q) = 0.9 * 1 - 0.1 * 0.98 = 0.802$$
$$\text{Soil (iwd)} = 2.85 + 0.98 = 3.83$$

So the path will be: A–B–F–M–Q.

Now, the calculations for the second iteration for the IWD are shown below. The water drop starts gain from the root of the node, i.e., node *A*.

Iteration 2:

$$P (A, B) = 13.41/28.41 = 0.472$$
$$P (A, C) = 1/28.41 = 0.351$$
$$P (A, D) = 9/28.41 = 0.316$$
$$P (A, E) = 5/28.41 = 0.1759$$

Since probability of node B is more we can traverse to node B.

$$\text{Time } (A, B) = \max\{(17 - 7), 1\}/\text{Velocity}(V_{\text{iwd}})$$
$$= 10/100 = 0.1 \text{ s}$$
$$\Delta \text{ Soil } (A, B) = 1/(1 + 1 * 0.1) = 0.9$$
$$\Delta \text{ Velocity } (A, B) = 100 + 1/1 + 1 * 13.41 = 100.069$$
$$\text{Updated Soil } (A, B) = 0.9 * 13.41 - 0.1 * 0.9 = 11.979$$
$$\text{Soil (iwd)} = 0.9 + 3.83 = 4.73$$

So the path will be: A–B.

$$P (B, F) = \text{Soil } (B, F)/\{\text{Soil } (B, F) + \text{Soil } (B, G)\}$$
$$= 8.003/11.003 = 0.727$$
$$P (B, G) = 0.272$$

As the probability of node F is greater so the water drop will flow to node F.

$$\text{Time } (B, F) = \{7 - 4\}/100.069 = 0.029$$
$$\Delta \text{ Soil } (B, F) = 1/(1 + 1 * 0.029) = 0.971$$
$$\Delta \text{ Velocity } (B, F) = 100.069 + 1/1 + 1 * 8.003 = 100.180$$
$$\text{Updated Soil } (A, B) = 0.9 * 8.003 - 0.1 * 0.971 = 7.1056$$
$$\text{Soil (iwd)} = 4.73 + 0.970 = 5.701$$

So the path will be: A–B–F.

Similarly, after reaching node F the water drop has to decide the next node. The calculations for the various parameters are shown below:

$$P(F, M) = 1/5.402 = 0.814$$
$$P(F, L) = 1/5.402 = 0.185$$

As the probability of node M is greater so the water drop will flow to node M

$$\text{Time}(F, M) = \{4 - 2\}/100.180 = 0.0199\,\text{s}$$
$$\Delta\,\text{Soil}(F, M) = 1/(1 + 1 * 0.0199) = 0.9804$$
$$\Delta\,\text{Velocity}(F, M) = 100.180 + 1/1 + 1 * 4.402 = 100.365$$
$$\text{Updated Soil}(F, M) = 0.9 * 4{,}402 - 0.1 * 0.98 = 3.863$$
$$\text{Soil(iwd)} = 5.701 + 0.98 = 6.814$$

So the path will be: A–B–F–M.
Now,

$$\text{Time}(M, R) = \{2 - 0\}/100.365 = 0.0199\,\text{s}$$
$$\Delta\,\text{Soil}(M, R) = 1/(1 + 1 * 0.0199) = 0.98$$
$$\Delta\,\text{Velocity}(M, R) = 100.365 + 1/1 + 1 * 1 = 100.865$$
$$\text{Updated Soil}(M, R) = 0.9 * 1 - 0.1 * 0.98 = 0.802$$

So the path will be: A–B–F–M–R.
In a similar manner, the algorithm is applied to compute various paths. The path with the maximum weight indicates the most trustworthy path.

Path 1: $A \rightarrow B \rightarrow F \rightarrow M \rightarrow Q$ (Total weight of path is: 30)
Path 2: $A \rightarrow B \rightarrow F \rightarrow M \rightarrow R$ (Total weight of path is: 26.617)
Path 3: $A \rightarrow B \rightarrow F \rightarrow L$ (Total weight of path is: 22.9476)
Path 4: $A \rightarrow B \rightarrow G \rightarrow N$ (Total weight of path is: 17.9881)
Path 5: $A \rightarrow D \rightarrow H \rightarrow O$ (Total weight of path is: 15)
Path 6: $A \rightarrow D \rightarrow H \rightarrow P$ (Total weight of path is: 9.6476)
Path 7: $A \rightarrow D \rightarrow I$ (Total weight of path is: 7.99)
Path 8: $A \rightarrow E \rightarrow J$ (Total weight of path is: 6)
Path 9: $A \rightarrow E \rightarrow K$ (Total weight of path is: 5.21)
Path 10: $A \rightarrow C$ (Total weight of path is: 3.88)
This data can be shown in Table 10.1.

The proposed algorithm uses IWD concept to generate all the possible independent paths with their priority, which is depicted in the path strength in Table 10.4. The weight of each path is calculated by adding the soil on all the links

Table 10.4 Path strength table

Path No	Path	Weight	Priority
1	A → B → F → M → Q	30	1
2	A → B → F → M → R	26.617	2
3	A → B → F → L	22.9476	3
4	A → B → G → N	17.9881	4
5	A → D → H → O	15	5
6	A → D → H → P	9.6476	6
7	A → D → I	7.99	7
8	A → E → J	6	8
9	A → E → K	5.21	9
10	A → C	3.88	10

of the path. It shows that the path with highest weight has the highest priority. The beauty of the algorithm is that it traverses all the paths as well as the nodes and computes the longest and the best path. Weight can be said as the trust value of that path and $\Delta Soil\ (i, j)$ is the trust of a particular communication between node i and j.

10.11   Summary

Wireless Sensor Network (WSN) is a network based on multiple low data rate and energy constrained sensor nodes. The purpose of the network is to monitor physical and environmental parameters such as temperature, pressure etc. The network consists of sensor nodes and gateway nodes, where the sensor nodes acquire the parameters and relay these cooperatively through other sensor nodes to the gateway nodes, which process and forward the data to the end user. Security in WSN is important since it not only provides reliability to the system but also helps in sustaining the network. Security threats can be introduced in WSN through various means, where the ongoing data transmissions can be tampered or the nodes can be altered to behave in an unpredictable manner. The objective of the case study was to present a review of various attacks and strategies used in overcoming the attacks. It discusses the security measures that can be implemented to detect the fraudulent nodes through trust and reputation models. Finally, comparative analysis is shown for the trust models discussed in the case study. Although research efforts have been made in regard to various approaches in trust and reputation model, there are still some challenges like high computational complexity pertaining to hardware implementation of WSN secure algorithms; memory; energy consumption and reliability of the overall WSN system, which need to be addressed. Since the sensors are characterized by constraints of energy, computation capability and memory, the design of the security services should be simple yet effective with regard to fraudulent node detection in WSN. Managing trust in Wireless Sensor Networks (WSNs) is a crucial issue and necessary steps should be taken for security purpose in an efficient, accurate and robust way. Providing this management would notably increase the security in such a sentient environment, supporting thus its development and deployment. Thus, bio-inspired approaches not only are less complex but are very efficient. The present case study uses bio-inspired approaches such as swarm intelligence, immune system and intelligent water drops are described.

References

Agah, A., Das, S. K., & Basu, K. (2004). A game theory based approach for security in wireless sensor networks. In *2004 IEEE International Conference on Performance, Computing, and Communications* (pp. 259–263).

Agah, A., Basu, K., & Das, S. K. (2005). Preventing DoS attack in sensor networks: a game theoretic approach. In *2005 IEEE International Conference on Communications, 2005, ICC 2005* (Vol. 5, pp. 3218–3222).

Agah, A., & Das, S. K. (2007). Preventing DoS attacks in wireless sensor networks: A repeated game theory approach. *International Journal Network Security, 5*(2), 145–153.

Agah, A., Basu, K., & Das, S. K. (2006). Security enforcement in wireless sensor networks: A framework based on non-cooperative games. *Pervasive and Mobile Computing, 2*(2), 137–158.

Aggarwal, K., Goyal, M., & Srivastava, P. R. (2012). Code coverage using intelligent water drop. *International Journal Of Bio-Inspired Computation, 4*(6), 392–402.

Ahmed, N., Kanhere, S. S., & Jha, S. (2005). The holes problem in wireless sensor networks: A survey. *ACM SIGMOBILE Mobile Computing and Communications Review, 9*(2), 4–18.

Alpcan, T., & Basar, T. (2006). An intrusion detection game with limited observations. In *12th International Symposium on Dynamic Games and Applications, Sophia Antipolis, France*.

Ng, A. (2012a). Retrieved February, 2013, from http://cs229.stanford.edu/notes/cs229-notes1.pdf.

Ng, A. (2012b). Video Lectures on Machine Learning. Retrieved February, 2013.

Atakli, I. M., Hu, H., Chen, Y., Ku, W. S., & Su, Z. (2008). Malicious node detection in wireless sensor networks using weighted trust evaluation. In *Proceedings of the 2008 Spring Simulation Multiconference* (pp. 836–843). Society for Computer Simulation International.

Auralius. (2009). Retrieved May, 2013 from https://decibel.ni.com/content/docs/DOC-5381.

Baljak, V., Tei, K., & Honiden, S. (2012). Classification of faults in sensor readings with statistical pattern recognition. In *SENSORCOMM 2012, The Sixth International Conference on Sensor Technologies and Applications* (pp. 270–276).

Bhaskaran, K., Triay, J., & Vokkarane, V. M. (2011). Dynamic anycast routing and wavelength assignment in WDM networks using ant colony optimization (ACO). In *2011 IEEE International Conference on Communications (ICC)* (pp. 1–6). IEEE.

Bhanot, S. (2008). Artificial Neural Network. In *Process control principle applications*.

Boukerch, A., Xu, L., & El-Khatib, K. (2007). Trust-based security for wireless ad hoc and sensor networks. *Computer Communications, 30*(11), 2413–2427.

Chen, H., Wu, H., Hu, J., & Gao, C. (2008, June). Event-based trust framework model in wireless sensor networks. In *International Conference on Networking, Architecture, and Storage, 2008. NAS'08* (pp. 359–364). IEEE.

Chen, H., Wu, H., Zhou, X., & Gao, C. (2007). Agent-based trust model in wireless sensor networks. In *Eighth ACIS International Conference on Software Engineering, Artificial Intelligence, Networking, and Parallel/Distributed Computing, 2007, SNPD 2007* (Vol. 3, pp. 119–124).

Chen, Y., Zhong, Y., Shi, T., & Liu, J. (2009). Comparison of two fitness functions for GA-based path-oriented test data generation. In *ICNC'09. Fifth International Conference on Natural Computation, 2009* (Vol. 4, pp. 177–181).

Chen, Z., Qiu, Y., Liu, J., & Xu, L. (2011). Incentive mechanism for selfish nodes in wireless sensor networks based on evolutionary game. *Computers & Mathematics with Applications, 62*(9), 3378–3388.

Cormen, T. H. (2009). *Introduction to algorithms*. Cambridge: MIT press.

Curiac, D. I., Volosencu, C., Doboli, A., Dranga, O., & Bednarz, T. (2007). Discovery of malicious nodes in wireless sensor networks using neural predictors. In *WSEAS Transactions on Computers Research* (Vol. 2, pp. 38–43).

Dargie, W. W., & Poellabauer, C. (2010). *Fundamentals of wireless sensor networks: Theory and practice*. New Jersey: John Wiley & Sons.

Dibrov, B. F., Livshits, M. A., & Volkenstein, M. V. (1977a). Mathematical model of immune processes. *Journal of theoretical biology, 65*(4), 609–631.

Dibrov, B. F., Livshits, M. A., & Volkenstein, M. V. (1977b). Mathematical model of immune processes: II. Kinetic features of antigen—Antibody interrelations. *Journal of theoretical biology, 69*(1), 23–39.

Dorigo, M., & Gambardella, L. M. (1997). Ant colony system: A cooperative learning approach to the traveling salesman problem. *IEEE Transactions on Evolutionary Computation, 1*(1), 53–66.

Du, W., Deng, J., Han, Y. S., & Varshney, P. K. (2006). A key predistribution scheme for sensor networks using deployment knowledge. *IEEE Transactions on Dependable and Secure Computing, 3*(1), 62–77.

Eschenauer, L., & Gligor, V. D. (2002). A key-management scheme for distributed sensor networks. In *Proceedings of 9th ACM conference on Computer and communications security*, pp. 41–47.

Fowler, A. C. (1981). Approximate solution of a model of biological immune responses incorporating delay. *Journal of Mathematical Biology, 13*(1), 23–45.

Ganeriwal, S., Balzano, L. K., & Srivastava, M. B. (2008). Reputation-based framework for high integrity sensor networks. *ACM Transactions on Sensor Networks (TOSN), 4*(3), 15.

Großschädl, J., Szekely, A., & Tillich, S. (2007). The energy cost of cryptographic key establishment in wireless sensor networks. In *Proceedings of the 2nd ACM Symposium on Information, Computer and Communications Security* (pp. 380–382). New York: ACM.

Horowitz, E., & Sahni, S. (1978). *Fundamentals of computer algorithms* (p. 206). Cambridge: Computer Science Press.

Jerome, J., Aravind, A. P., Arunkumar, V. & Balasubramanian, P. (2005). LabVIEW based intelligent controllers for speed regulation of Electric Motor. In *Proceedings of the IEEE on Instrumentation and Measurement Technology Conference, 2005* (Vol. 2, pp. 935–940).

Jonker, C. M., & Treur, J. (1999). Formal analysis of models for the dynamics of trust based on experiences. In *Multi-Agent System Engineering* (pp. 221–231). Berlin: Springer.

Jøsang, A., Ismail, R., & Boyd, C. (2007). A survey of trust and reputation systems for online service provision. *Decision Support Systems, 43*(2), 618–644.

Joseph, J. F. C., Lee, B. S., Das, A., & Seet, B. C. (2011). Cross-layer detection of sinking behavior in wireless ad hoc networks using SVM and FDA. *IEEE Transactions on Dependable and Secure Computing, 8*(2), 233–245.

Kayalvizhi, R., Vijayalakshmi, M., & Vaidehi, V. (2010). Energy analysis of RSA and ELGAMAL algorithms for wireless sensor networks. In *Recent Trends in Network Security and Applications* (pp. 172–180). Berlin: Springer.

Kim, T. K., & Seo, H. S. (2008). A trust model using fuzzy logic in wireless sensor network. *World Academy of Science, Engineering and Technology, 42*(6), 63–66.

Lee, I. (2007). Software System Lecture Note: Security. Retrieved August 28, 2014, from http://www.cis.upenn.edu/lee/07cis505/Lec/lec-ch9asecurity-v2.pdf.

Li, X., & Lyu, M. R. (2008). A novel coalitional game model for security issues in wireless networks. In *Global Telecommunications Conference, 2008, IEEE GLOBECOM 2008* (pp. 1–6). IEEE.

Lopez, J., Roman, R., Agudo, I., & Fernandez-Gago, C. (2010). Trust management systems for wireless sensor networks: Best practices. *Computer Communications, 33*(9), 1086–1093.

Luo, J., Liu, X., & Fan, M. (2009). A trust model based on fuzzy recommendation for mobile ad-hoc networks. *Computer Networks, 53*(14), 2396–2407.

Lupu, T. G. (2009). Main types of attacks in wireless sensor networks. In I. Rudas, M. Demiralp, & N. Mastorakis (Eds.), *WSEAS International Conference, Proceedings, Recent Advances in Computer Engineering* (Vol. 9).

Raj, M. R. C., Kumar G. E. P., Kusampudi, K. (2013). A survey on detecting selfish nodes in wireless sensor networks using different trust methodologies. *International Journal of Engineering and Advanced Technology (IJEAT), 2*(3), 197–200.

Mármol, F. G., & Pérez, G. M. (2012). TRIP, a trust and reputation infrastructure-based proposal for vehicular ad hoc networks. *Journal of Network and Computer Applications, 35*(3), 934–941.

Mármol, F. G., & Pérez, G. M. (2011). Providing trust in wireless sensor networks using a bio-inspired technique. *Telecommunication systems, 46*(2), 163–180.

Meng, Y., & Li, W. (2013). Evaluation of detecting malicious nodes using Bayesian Model in wireless intrusion detection. In *Network and System Security* (pp. 40–53). Berlin: Springer.

Momani, M., & Challa, S. (2010). Survey of trust models in different network domains. *arXiv preprint* arXiv:1010.0168.

Momani, M., Challa, S., & Alhmouz, R. (2008). BNWSN: Bayesian network trust model for wireless sensor networks. In *Mosharaka International Conference on Communications, Computers and Applications, 2008. MIC-CCA 2008* (pp. 110–115).

Muller, K., Mika, S., Ratsch, G., Tsuda, K., & Scholkopf, B. (2001). An introduction to kernel-based learning algorithms. *IEEE Transactions on Neural Networks, 12*(2), 181–201.

Ni, K., Ramanathan, N., Chehade, M. N. H., Balzano, L., Nair, S., Zahedi, S., et al. (2009). Sensor network data fault types. *ACM Transactions on Sensor Networks (TOSN), 5*(3), 25.

Neuman, B. C., & Ts' O. T. (1994). Kerberos: An authentication service for computer networks. *Communications Magazine IEEE, 32*(9), 33–38.

Newsome, J., Shi, E., Song, D., & Perrig, A. (2004). The sybil attack in sensor networks: analysis & defenses. In *Proceedings of the 3rd international symposium on Information processing in sensor networks* (pp. 259–268).

Pathan, A. S. K., Lee, H. W., & Hong, C. S. (2006). Security in wireless sensor networks: issues and challenges. In *The 8th International Conference on Advanced Communication Technology, 2006. ICACT 2006* (Vol. 2, p. 6).

Pehr, S. (2008). An Analysis of WSN Security Management. *Master of Science Thesis*, Stockholm, Sweden, Chapter-2, pp. 6–12.

Rathore, H., & Jha, S. (2013). Bio-inspired machine learning based wireless sensor network security. In *2013 World Congress on Nature and Biologically Inspired Computing (NaBIC)* (Vol. 5, pp. 140–146).

Rathore, H., Badarla, V., Jha, S., & Gupta, A. (2014). Novel approach for security in wireless sensor network using bio-inspirations. In *Proceedings of IEEE International Conference on Communication Systems and Networks (COMSNETS)* (Vol. 6, pp. 1–8).

Reddy, Y. B. (2012). Trust-based approach in wireless sensor networks using an agent to each cluster. *International Journal of Security, Privacy and Trust Management, 1*(1), 19–36.

Reddy, Y. B. (2009). A game theory approach to detect malicious nodes in wireless sensor networks. In *Proceedings of International Conference on Sensor Technologies and Applications (SENSORCOMM)* (Vol. 3, pp. 462–468).

Resnick, P., Kuwabara, K., Zeckhauser, R., & Friedman, E. (2000). Reputation systems. *Communications of the ACM, 43*, 45–48.

Ringwald, M., & Romer, K. (2007). Deployment of sensor networks: Problems and passive inspection. In *Proceedings of Intelligent Solutions in Embedded Systems* (Vol. 5, pp. 179–192).

Scholkopf, B., & Smola, A. J. (2001). Learning with kernels: Support vector machines, regularization, optimization, and beyond (pp. 204–205). Cambridge: MIT Press.

Shah-Hosseini, H. (2007). Problem solving by intelligent water drops. *In IEEE Congress on Evolutionary Computation, 1*, 3226–3231.

Shah-Hosseini, H. (2009). *Optimization with the nature-inspired intelligent water drops algorithm*. INTECH Open Access Publisher.

Sharma, R., Chaba, Y., & Singh, Y. (2010). Analysis of security protocols in wireless sensor network. *International Journal of Advanced Networking and Applications, 2*(3), 707–713.

Sharma, K., & Ghose, M. K. (2010). Wireless sensor networks: An overview on its security threats. *International Journal of Computer Applications Special Issue on Mobile Ad-hoc Networks.*

Shigen, S., Yue, G., Cao, Q., & Yu, F. (2011). A survey of game theory in wireless sensor networks security. *Journal of Networks, 6*(3), 521–532.

Soderman, P. (2008). An analysis of wsn security managemant. *Master of Science Thesis.*

Soliman, H. H., Hikalb, N. A., & Sakrb, N. A. (2012). A comparative performance evaluation of intrusion detection techniques for hierarchical wireless sensor networks. *Egyptian Informatics Journal, 13*(2), 225–238.

Suykens, J. A., & Vandewalle, J. (1999). Least squares support vector machine classifiers. *Neural Processing Letters, 9*(3), 293–300.

Vapnik, V. (2000). *The nature of statistical learning theory*. New York: Springer.

Wang, Y., Attebury, G., & Ramamurthy, B. (2006). A survey of security issues in wireless sensor networks. *CSE Journal Articles, 8*(2).

Wang, W., Chatterjee, M., & Kwiat, K. (2009). Coexistence with malicious nodes: A game theoretic approach. In *Proceedings of International Conference on Game Theory for Networks (GameNets 09)* (pp. 277–286).

Wagstaff, K., Cardie, C., Rogers, S., & Schrödl, S. (2001). Constrained k-means clustering with background knowledge. In *ICML* (Vol. 1, pp. 577–584).

Wikipedia. (2013). Retrieved December 3, 2013, from http://en:wikipedia:org=wiki= Computationalcomplexityofmathematicaloperations.

Zia, T., Zomaya, A., Ababneh, N. (2007). Evaluation of overheads in security mechanisms in wireless sensor networks. In *Proceedings of International Conference on Sensor Technologies and Applications* (pp. 181–185).

Chapter 11
Bio-inspired Approaches in Various Engineering Domain

Abstract Nature has always inspired scientific research to flourish in one way or other. Nature is one of those greatest expansions that have shown extraordinary results in every respect. The biological activities that are occurring around us have splendid and magnificent characteristics that have inspired us in many ways. For instance, ants coordinate themselves to perform foraging; birds synchronize themselves to show beautiful patterns; fireflies show impressive patterns of luminosity at night, etc. Therefore, engineers and scientists of diverse domains are engaging themselves in instigating innovative design architectures in resolving different difficulties and challenges. This chapter deals with providing elaborative details on bio-inspired research in streams, such as energy, agriculture, aerospace, electrical, mechanical.

11.1 Bio-inspired Energy Systems

Nature's biological systems have caught the attention of scientific community of energy systems. Energy systems have started taking attention from biological systems to improve current technologies in energy domain. This section points out the on-going or held research on bio-inspired energy systems. Following are few examples of such work.

11.1.1 Bio-inspired Solar Energy Program at CIFAR (Canadian Institute for Advance Research)

Solar energy is radiant light and heat from the Sun which can be harnessed into useful form through a range of efficient technologies routes, such as solar thermal, solar photovoltaics, solar architecture, and solar greenhouse, etc. Bio-inspired solar energy system takes motivation from photosynthesis, which has advanced over billions of years to productively change over sun-oriented vitality to electrical

© Springer International Publishing Switzerland 2016
H. Rathore, *Mapping Biological Systems to Network Systems*,
DOI 10.1007/978-3-319-29782-8_11

and chemical energy (Akerlof 2015). Specialists in the system are utilizing these lessons to change over photons into electrons all the more proficiently, and to make new fuels specifically from sun powered vitality. Goal of the project is to create proficient methods to reap the vitality from photons. Chlorophyll, a vital component in photosynthetic creatures has upgraded the procedure in various ways. For instance, it sorts out itself into varieties of receiving wires that work in catching the vitality from a photon and transporting it to a response focus, where the vitality is utilized to raise an electron to a high vitality state. Lessons from these reception apparatus clusters could make counterfeit sun powered catch significantly more effective.

Another significant example is in catalysis and fuel storage. Plants use sun oriented vitality to change over carbon dioxide to fuel sugars with unbelievable proficiency. CIFAR is attempting to convey the same effectiveness to change over sun powered energy to chemical energy, which could be put away, and later changed over to electrical vitality.

11.1.2   Bio-inspired Optimization of Sustainable Energy Systems

Global warming and fossil fuel depletion increasingly place the development of sustainable energy systems around the globe. Major investments in new renewable energy technologies and systems to improve energy efficiency and reduce greenhouse gas emissions will continue to grow in the coming decades. Zheng et al. (2013) summarized the recent advances in bio-inspired optimization methods, including artificial neural networks, evolutionary algorithms, swarm intelligence, and their hybridizations, which are applied to the field of sustainable energy development.

11.1.3   Biomimicry Innovations for Energy Sustainability

Sustainable energy is the form of energy obtained from non-exhaustible renewable energy resources, such as solar energy, wind energy, bio energy, tidal energy, wave energy, geothermal energy, hydroelectricity energy designed to improve energy efficiency. Honey and Pagani (2013) provides insights on biomimicry, nature's plans technological advancements with examination, which offers guarantee for "bio-inspired energy" to make more proficient energy production, energy storage, and energy delivery with developments that duplicate the outlines of common frameworks. Opportunity exists for many-sided quality sciences and system models that exist in nature to propel current innovations, base, and strategy approaches in the energy sector. Many-sided quality sciences use techniques for assisting us with comprehension appropriated for energy systems and savvy framework elements with bidirectional correspondence. They can likewise give models and rules to educate energy improvements, rising practices, and future difficulties. In 2012, the Joint Research Center in the EU started a vast scale, worldwide push to coordinate these ideas of intricacy into keen matrices and energy conveyance. As a subsegment of this unpredictable frameworks way to deal with taking care of future energy requests, it is theorized that energy systems for individuals will profit by biomimicry exploration to invigorate new thoughts for energy innovation developments, proficient lattice outlines, and systems for ideal energy conveyance in the midst of a dynamic situation. The multidisciplinary endeavors of the multifaceted nature methodology can help in handling the energy-related issues from a few perspectives and conquering any hindrance to incorporate ideas far from the building viewpoint, i.e., biological inspired solutions in the future energy solutions. Some of the ventures which energize biomimicry research for energy are described below:

(a) **Energy production and harvesting (e.g., photosynthesis inspired technologies)**:
    The work focuses on artificial photosynthesis, wind turbine planning, black butterfly wing solar collection structure, marine, and wind turbines that mimic non-smooth surface heliostat positioning for solar plants from sunflowers.
(b) **Energy storage (e.g., nature inspired battery technologies)**:
    Conductors for batteries, hydrogen production happens through hierarchical structure in butterfly wing.
(c) **Energy delivery and grid efficiency (e.g., efficient natural networks for grid design)**
    More efficient led light is produced by taking inspiration from fireflies and threats communication mechanism from insects.

11.1.4 Bio-inspired Artificial Light-Harvesting Antennas for Enhancement of Solar Energy

Light absorption is the essential stride in any photovoltaic gadget, subsequently panchromatic light gathering is a basic condition to amplify the effectiveness of a sun-based cell. In fact, the photocurrent thickness is specifically corresponding to the light division consumed by the cell in respect to the entire approaching sunlight-based flux. By impersonating the light collecting radio wires of characteristic photosynthetic frameworks it is conceivable to improve the light catch of photoanodes in dye-sensitized solar cells (DSCs) by utilizing distinctive engrossing units which channel the occurrence of light to the sensitizer by energy exchange. In this point of view (Odobel et al. 2013) presents bio-propelled methods to enhance the light gathering effectiveness of color sharpened DSCs by the radio wire. The work presents the synopsis of the absolute most critical and late advancements in receiving wires usage impact to enhance the photocurrent thickness in DSCs. The work highlights how new and imaginative multi-chromophoric sensitizers can adequately expand and upgrade the assimilation cross-area which empowers to deliver higher photocurrent thickness both in fluid and strong state sun powered cells.

Above-mentioned are few research zones of bio-inspired energy systems where nature solves problem in most optimized possible way. Therefore, taking cue from biosystems to meet energy needs seems quite optimistic and to establish the same few examples of conducted or ongoing research on bio-inspired energy systems have been mentioned in the section above.

11.2 Bio-inspired Agriculture Systems

An agricultural system is an integration of components and system essentially required within a prescribed boundary to achieve a specified production objective on behalf of the system. Engineering interventions in agriculture have become imperative to improve productivity, reduce the cost of production, drudgery, and improve livelihood opportunities including enhanced income. Farm mechanization for timeliness of operations, processing technologies for loss reduction and value addition, conservation of natural resources, and energy management are some of the core areas in which the benefits of engineering interventions have begun to be felt (Singh et al. 2013). The present section provides details on bio-inspired research in agriculture.

11.2.1 CO–CH Project

The reasoning of bio-motivated frameworks comprises on concentrating on natures innovations, traps, and stratagems to create designing arrangements supplied with life like properties. The CO–CH project utilizes bio-motivated computational strategies which are fit for creating complex models to anticipate/depict the site specific behavior of crops (Barreto et al. 2006).

Bio-inspired techniques have turned out to be capable devices for demonstrating and forecast utilizing numerical noisy data and nonnumerical data. They seem to perform better without strong assumptions on the information. Certain bio-inspired techniques give inventive approaches to prepare and picture high-dimensional data, for instance, fuzzy logic formalism empowers the mix of uncertain data, and the generation of human understandable outputs. It can process any data (numerical/noisy/nonnumerical), train, test, and model it to predict the explanation and visualization of data. Construction of models using growing data bases is a challenging issue. In CO–CH project, following $3i$ methodologies are taken into account for model validation and exploitation:

(a) *Incremental Modeling*: The information of fruit crops will be constantly gathered along with the demonstrating procedure. Therefore, the model must have the capacity to adjust its parameters as indicated by the progressions of approaching data. This procedure is firmly identified with nonstop web learning frameworks in which the model structure must be plastic yet sufficiently stable to have the capacity to learn new qualities of information while holding past data.

(b) *Integration of heterogeneous information*: The modeling methodology needs to take into consideration the likelihood to incorporate data assembled from various sources. The project is intrigued to incorporate master learning, customary learning, and data acquired by agronomical and atmosphere investigation. Keeping in mind the end goal to manage various wellsprings of different nature of information, the project propose to build up an alleged "*blend of specialists*" methodology.

(c) *Intelligent visualization*: Building a model is seldom an end in itself; rather, the objective of most investigation is to settle on a choice. To help with this examination, the project propose the improvement of keen interfaces that take into account visual choice based on bio-inspired techniques and data mining.

11.2.2 Bio-inspired Sensing for Agriculture Robots

Robotics in agriculture turns into a more practical in the most recent 10 years, particularly with the rise of minimal effort gadgets, sensors for direction and light weight high power batteries. Keeping in view the end goal to capacity in an unorganized environment of a rural field or a greenhouse a robot must be self-sufficient with its own method in the view of the earth. Current advances depend on vision and profundity estimation by Ladar or stereo vision. Albeit ultrasonic separation estimation is utilized as a part of indoor robots which is not regular in the field of mechanical autonomy. This is on the grounds that the echoes are not vigorous and dependable. Bats utilize effectively in significantly more troublesome circumstances then a farming robot. The work copies the bats' capacity to make ultrasonic signs and their capacity to handle the arrival eco with a specific end goal to control farming robots (Kosa 2015).

AGRY-bot is a wheeled portable robot for crop estimation that is equipped for exploring in a greenhouse and giving a quantitative measure to the measure of peppers on the plants. The work utilizes the handling systems for propulsion by the bat's sonar. The work is likely to expand these introductory results to different harvests and distinctive portable stages.

11.2.3 Bio-inspired Special Wettability

Nature is a school for researchers and designers. After four and a half billion years of stringent development, a few animals in nature display entrancing surface wettability. Biomimetics, copying nature for designing arrangements, gives a model to the advancement of practical surfaces with uncommon wettability. As of late, bio-inspired extraordinary wetting surfaces have pulled in wide experimental consideration for both key examination and down to earth applications, which has turned into an inexorably hot exploration theme. The Critical Review by Liu et al. (2010) outlines the recent work in bio-inspired uncommon wettability, with an emphasis on lotus leaf enlivened self-cleaning surfaces, plants and bugs propelled anisotropic super-hydrophobic surfaces, mosquito eyes roused super-hydrophobic antifogging coatings, creepy crawlies roused super-hydrophobic antireflection coatings, flower petals, and gecko feet motivated high glue super-hydrophobic surfaces, bio-enlivened water gathering surfaces, and superlyophobic surfaces, with specific spotlight in the most recent 2 years.

11.2.4 Biorefinery: A Bio-inspired Process to Bulk Chemicals

Biorefinery is a multidisciplinary and complex concept responsible for production of value-added bio products, such as chemical building blocks, material, etc. and bio energy from biomass, such a bio fuels, bio gas and producer gas for power and heat, etc. (Sanders et al. 2007) portrays a few samples of information concentrated innovations for the creation of chemicals from biomass, which exploit the biomass structure in a more effective manner than the generation of powers or power alone. The consumption in fossil feedstocks, expanding oil costs, and the biological issues connected with CO_2 emissions are compelling the improvement of option assets for vitality, transport fills, and chemicals. For instance, the supplanting of fossil assets with CO_2 neutral biomass is an example. Partnered with this, is the change of unrefined petroleum items which uses essential items (ethylene) and their trans-formation into either materials or (useful) chemicals with the guide of co-reagents. Major examples include smelling salts, by different procedure ventures to present functionalities, i.e., $-NH_2$ into the basic structures of the essential items. Furthermore, numerous items found in biomass regularly contain functionalities. Along these lines, it is alluring to adventure this so as to bypass the utilization and planning of co-reagents and to take out different procedure ventures by using suitable biomass-based forerunners for the creation of chemicals.

11.3 Bio-inspired Aerospace Systems

The developing utilization of bio-inspired structures in aviation and numerous different disciplines has emerged from their high particular quality and solidness, and simple (low vitality) creation forms when contrasted with more traditional materials. Also, human has the capacity to shape and tailor their structure to deliver all the more efficiently productive basic setups. There is a developing yearning in the business to comprehend these classes of bio-inspired materials for interdisciplinary outline work and to tailor these exotic materials to address the issues of industry application. Looking into the behavior of bio-inspired structures properties at different levels (MIT 2009) as listed below:

- Material level (e.g., micro-cracking, failure theories, plasticity, and fracture)
- Structural (e.g., torsion, joints, biaxial bending)
- Element level (e.g., failure modes, shear transfer, creep, and ductility)
- Systems level (e.g., morphing, IUAV, spacecraft, cardiovascular, bone, etc.)

The goal of bio-inspired frameworks in the field of aviation frameworks is to apply bio-inspired nanomaterials, insightful, and transforming structures standards for an innovative outline arrangement, e.g. bio-inspired energy structures. It helps in comprehending the structure, properties, and execution at diverse length scales with respect to the various leveled association of bio-inspired structures, for example, bone. Biologically inspired designing materials that show mechanical properties expand the quality and durability for a material in respect to its constituents. It helps in comprehending the utilization of human components in configuration, building and to do interdisciplinary outline work utilizing these bio-motivated structures.

11.3.1 Biomorphic Explorers

Biomorphic explorers are among the biologically inspired aviation frameworks that the NASA Jet Propulsion Laboratory is concentrating on Noor et al. (2000). These little, committed, ease gadgets catch a percentage of the natural wayfarers, incorporating reconfigurable units with adaptable versatility (for example, airborne, surface, and subsurface explorers). They may likewise highlight control by versatile, flaw tolerant, bio-inspired calculations to match evolving encompassing/landscape conditions.

These elements empower far reaching investigation at lower expense with more extensive scope through agreeable association of landers, meanderers, and a mixed bag of reasonable low-mass biomorphic explorers, for example, biomorphic lightweight flyers, balloons, and powered aircraft. Particular science goals focused for biomorphic explorer missions incorporate climatic data gathering by circulating numerous site estimations, close-up imaging for land site choice, arrangement of surface payloads (for example, instruments, surface analyses, and test return surveillance). Different samples of biological structures that use composite materials and cements are given in Renton (2001).

11.3.2 Bio-inspired Design of Aerospace Composite Joints for Improved Damage Tolerance

The work presented by Burns et al. (2012) utilizes a bio-inspired configuration procedure in view of tree limb joints to enhance the harm resilience of co-cured composite T-joints. The configuration of tree limb joints at diverse length scales from the microstructural to the large scale length scale was explored. X-beam processed tomography of a pine tree uncovered three fundamental elements of tree limb joints which give high auxiliary proficiency and harm resilience:

- Coordinated outline with the branch installed into the focal point of the trunk
- Three-dimensional fibril lay-up in the central stress directions;
- Variable fibril thickness to accomplish isostrain conditions through the joint association.

Exploration introduced in this work adjusts the inserted auxiliary element of tree joints into a carbon/epoxy T-joint. The spine employs were installed to 25, 50, and 75 % of the skin's profundity of the composite T-joint to impersonate the configuration of tree limb joints. Exploratory testing uncovered that the bio-propelled T-joint outline with coordinated adherents had expanded standardized inelastic strain vitality (characterized as ductility), expanded standardized consumed strain vitality to disappointment, and higher burden conveying limit taking after harm start (harm resistance) contrasted with an ordinarily reinforced T-joint. Be that as it may, these changes were accomplished to the detriment of prior onset of harm start in the T-joints.

11.3.3 Pneumatic Artificial Muscles

Pneumatic artificial muscles (PAMs) were initially brought about by Gaylord and Eirich (1950) and have been explored for utilization in prosthetic and mechanical gadgets, delicate **applies** autonomy, transforming and nasty structures, and aviation applications. Recently, PAMs have been truly considered for aviation applications. PAMs have numerous alluring qualities for execution in these ranges. They are lightweight actuators that create elevated amounts of power and huge, usable stroke at moderate actuation pressures (<620 kPa) (Wereley 2015). The potential usage of PAM actuators broaden past their elite levels. PAMs are actually agreeable and are exceedingly tolerant to misalignment and hasty stacking. Air can be conveyed to PAMS by means of adaptable, lightweight, low weight tubing. PAMs are exceedingly agreeable to dispersed activation ideas. PAMS are utilized in aviation frameworks including space apply autonomy for controller and end effectors, EVA gloves, trailing edge folds in helicopter rotor edges and scaled down actuators for delicate robots.

11.4 Bio-inspired Electrical Systems

Bio-inspired electrical systems aim for developing methodologies coupling application of the scientific principle for the improved industrial production. The production is in the areas of electric power production and utilization. The engineering lays the foundation of high technological development in circuit designing, analyzing radar signals, digital design, etc. The section presents some of the work in the field of bio-inspired developments in electrical systems.

11.4.1 Biologically Inspired Electrically Small Antenna Arrays with Enhanced Directional Sensitivity

Numerous crawlies have intense directional listening abilities and have the capacity to localize sound sources of interest with an astounding level of exactness. A similarity can be drawn between the sound-related frameworks of such creepy crawlies and electrically little reception antenna arrays that exhibit upgraded affectability to the bearing of landing of an electromagnetic wave, contrasted with regular arrays occupying the same aperture. Enlivened by this, the work presented by Behdad et al. (2011) examines the configuration of biologically inspired electrically small antenna arrays which exhibits the listening system of such crawlies. A strategy for planning such antenna arrays is introduced, and the tradeoffs included in accomplishing this improved affectability are examined. Reenactment and estimation consequences of two manufactured models are additionally displayed and talked about in this work.

11.4.2 Biologically Inspired At-scale Robotic Insect

Biology is a valuable apparatus when connected to designing difficulties that have been settled in nature. The work presented by Wood (2008), has the objective of making an insect estimated size, genuinely smaller scale air vehicle by investigating natural standards. These standards give experiences on the most proficient method to produce adequate push to maintain flight for centimeter-scale vehicles. Here, it is demonstrated how novel assembling standards empower the production of the mechanical and aeromechanical subsystems of a microrobotic gadget that is equipped for Diptera-like wing directions. The outcomes are a special microrobot: a 60 mg mechanical creepy crawly that can create adequate push to quicken vertically. Albeit still remotely controlled, this micromechanical gadget speaks to huge advancement toward the production of self-sufficient creepy crawly estimated miniaturized scale air vehicles.

11.4.3   Biomimetic and Bio-inspired Robotics in Electric Fish Research

Electric knife fish have charmed both scholars and specialists for a considerable length of time with their one of a kind electrosensory framework and spry swimming mechanics. Investigation of these fish has brought about models that enlighten the standards behind their electrosensory framework and interesting swimming capacities (Neveln et al. 2013). These models have revealed the components by which knifefish produce push for swimming forward and in reverse, floating, and hurling dorsally utilizing a ventral prolonged middle fin. Dynamic electrosensory models enlivened by electric fish take into account short proximity detecting in turbid waters where other detecting modalities come up short. Simulated electrosense is fit for helping route, recognition, and separation of objects and mapping the earth, all errands for which the fish use electrosense widely. While robotic ribbon fin and artificial electrosense exploration has been sought after independently to lessen complexities that emerge when they are joined, electric fish have succeeded in their natural specialty through close coupling of their detecting and mechanical frameworks. Future combination of electrosense and ribbon fin technology into a knifefish robot ought to bring about a vehicle fit for exploring complex 3D geometries inaccessible with flow submerged vehicles, and additionally give bits of knowledge into how to plan versatile robots that coordinate high data transmission detecting very responsive multidirectional movement.

11.4.4   Future Power Grid Inspired from Brain

The unmatched capacity of the human cerebrum to process and comprehend a lot of complex information has gotten the consideration of architects working in the field of control frameworks. Neuroscientists and architects are utilizing neurons developed as a part of a dish to control reenacted force frameworks. Concentrating on neural systems incorporates and reacts to complex data to rouse new techniques for dealing with the nation's constantly changing power supply and request (Thompson 2013). The objective is to decipher the physical and useful changes that happen as living neuronal system which learns from numerical comparisons, at last prompting a more brain like control framework. The examiners have effectively "taught" a living neuronal system how to react to complex information and have fused these discoveries into reenacted variants called bio-inspired artificial neural systems (BIANNS).

11.5 Bio-inspired Mechatronics Systems

It's high time to approach research in interdisciplinary manner for novel design paradigms in mechatronic systems. In fact, the same has been happening since quite a period now. One discipline is inspiring their counterparts in some marvelous innovations. So as biological systems has inspired energy systems, Computer networks, electrical systems, etc.

In this section a discussion on how biology has inspired mechatronic systems has been made. Examples of main research and innovations developed in this respect are explained below. Biological inspiration research in mechatronic systems can be classified in areas like actuation, sensing, locomotion and mechanisms, and control.

11.5.1 Bio-inspired Mechatronics

Animals are very maneuverable and agile creatures with power efficiency and possess robust motion on ground, water, air, walls, and sands. Crawling motion is quite efficient on wide ranges of complex lands while its cost of transport is relatively high due to high frictional cost of transport in and also the inertial cost of transport, which is proportional to the square of the motion speed during high-speed crawling motion. Snakes paved the way for the very first crawling robotic platform (Taylor 1974) by studying their locomotion biomechanics in 1974 since snakes can move efficiently on a wide range of terrains with large contact and are very strong under hostile environments due to sealed skin. Their ability to swim, jump, and climb using surface friction and glide have attracted researchers to design and build snake-inspired robots (Transeth et al. 2009; Hopkins et al. 2009; Shugen 2001; Ishikawa et al. 2010).

Fish has also inspired many robotic platforms. Firstly, in 1978 small-scaled automatic mechanical fish was developed in Japan. Also, the production of Robot Tuna in 1994 stimulated the extensive research interests in bio-inspired swimming robots (Triantafyllou and Triantafyllou 1995). Since then, many fish robots have been developed over the past 20 years (Zhou and Low 2012; Low 2009; Yu et al. 2012; Low and Willy 2006; Hu et al. 2006; Colgate and Lynch 2004; Liu and Hu 2010).

11.5.2 Bio-inspired Actuation

Skeletal muscles of vertebrates have properties such as high power density (up to 100 W/kg), high strain (up to 40 %), high stresses (up to 0.35 N/mm^2), high efficiency (up to 35 %), stiffness tuning capability (up to five times stiffness change), high strain rates (up to 5 lengths/s). Furthermore, they also have the capability to multifunction (e.g., can be used as a brake) and have high durability (up to billions of cycles), self-sensing capability, and self-repairing capability. All these properties are there in contractile actuators (motors) (Dawson and Taylor 1973). Hierarchical structure of muscles with fibers with parallel and distributed actuation architecture has helped in developing soft muscle-like actuators with similar properties. Such examples of muscle inspired actuators includes electroactive polymer actuators, conductive or ionic polymer actuators, shape memory alloys (SMAs), and piezoelectric actuator-based (Ishihara et al. 1996; Otsuka and Wayman 1999; Ueda et al. 2007).

11.5.3   Bio-inspired Control

Animals have this phenomenal ability of dealing with unstructured outdoor environments to exhibit skills which are impressive in terms of agility and efficiency. These skills are still far better than robots. For example a cat's ability of jumping, hunting, running, and climbing trees is yet to be implemented in robots. Experiments on these animals have shown that many of these behaviors exist at a low level in the vertebrate central nervous system, i.e., brainstem and the spinal cord (Whelan 1996; Bizzi et al. 2000; Grillner 2006). Their system is organized in such a way that their spinal circuits are responsible for generating rhythmic patterns and that higher level centers such as the motor cortex, cerebellum, and basal ganglia are responsible for modulating according to environmental conditions (Stein 1999). A key element in their system is the central pattern generators (CPGs) which are located in the spinal cord. CPGs are neural networks which are capable of producing coordinated patterns of rhythmic activity without any rhythmic inputs from sensory feedback and they can be easily activated and controlled by relatively simple signals from higher control centers (Delcomyn 1980). The circuits thus formed, acts or represents the basic building blocks. A conceptual image of these spinal cord circuits has been proposed where a marionette puppet on strings (Loeb 2001) and where a pulling a few strings, i.e., activating a few descending pathways is able to generate complex coordinated movements of entire limbs.

11.5.4   Future of Bio-inspired Mechatronics

Apart from the above-mentioned progress many fields are still to explore and some problems still remain unchallenged. Most of the work till now focused is on single locomotion while animals may have multimodal locomotion of adverse environments. Also, design and control systems for high-speed locomotion with phenomenal maneuverability are still a very challenging job. For example, flapping wing based flight is still to achieve. Also, bio-inspired robots are not multi terrain but animals cam move on variety of terrains.

But the way research is going on this area it gives a hope that all these limitations will soon be tackled by scientists.

11.6 Bio-inspired Civil Engineering

Taking lessons from nature and biology into civil engineering to add to another era of biologically well disposed, savvy answers for the advancement and recovery of versatile and maintainable common base frameworks is the present need of structural specialists. The fact of the matter is to move from the construction development generally concrete substantial, beast power way to deal with foundation, and supplant it with enhanced, effective, and maintainable answers for geotechnical rehearse. Specialists expand on nature's work in an assortment of routes, from creating bio-based strategies for fortifying soils to anticipate disintegration, and battle the dirt liquefaction amid tremors, to formulate innovations that match the tunneling abilities of creepy crawlies and little warm-blooded creatures (Fell 2015).

The scientists trust that such leaps forward will prompt the development of versatile streets, scaffolds, dams, force plants, pipelines and structures, alongside more proficient, and powerful asset recuperation operations.

11.6.1 Corrosion Protection via Application of Bacteria and Bio-polymers

The change of material properties of concrete-based materials, e.g., reduced penetrability, porosity; uniform dissemination of hydration items is of a critical significance having identified strength and supportability of concrete and fortified solid structures. With this regard, a late and novel methodology is to make self-mending frameworks that will have the capacity to recoup at least partly their initially planned execution. Bio-based cement-based materials were at that point demonstrated to have the limit of lessened break width and porousness separately as a consequence of bacterial action in the concrete grid. New methodologies, identified with erosion control for off shore structures, include the utilization of bacterial, hydrogen devouring living beings, bio-polymers, and nanocomposite or sol-gel covering for expanded administration life and enhanced imperviousness to microbiological consumption (MIC). The work introduced by Koleva et al. (2013) set up self-mending instrument in "bio-concrete" and detest the idea and methodology of as of late started research concerning MIC erosion control.

11.6.2   Nondestructive Evaluation of Civil Infrastructures

Concept of Civil infrastructure involves the design, analysis, and management of infrastructure supporting human activities including different utilities such as electric power, oil and gas, water and wastewater disposal, communications, transportation, etc. Evaluating the state of a structure is important to focus its security and unwavering quality. Ideally, structural health monitoring ought to be like medicinal health checking of the body. In medical health checking, the life signs, for example, heartbeat and pulse give a general sign of the general soundness of the body. This is comparable to worldwide structural health monitoring, in which harm to the structure can be recognized by measuring changes in the worldwide properties of the structure. When the body signs demonstrate a peculiarity, a battery of tests is done to focus the reason for the inconsistency. Comparably in basic structural health monitoring, nondestructive assessment (NDE) can be utilized to focus the nature of the damage. NDE routines to focus neighborhood harm are additionally turning out to be more acknowledged. The work displayed by Chang and Liu (2003) depicts a portion of the recent advancement of National Science Foundation ventures around the exploration.

11.6.3   Designing Model Systems for Enhanced Adhesion

Nature gives motivation to improved control of grip through various samples extending from geckos to bouncing bugs. The essential procedure in these cases is the incorporation of patterns, particularly high-perspective proportion topographic elements, to astutely boost bond powers while keeping up simplicity of discharge. Recently, significant exploration endeavors have been given toward the under-standing, advancement, and streamlining of manufactured analogs to these cases in nature. In this article, the work introduced by Chan et al. (2007) gives knowledge into the instruments that prompt improved control of interfacial properties through designing, the procedures that can be utilized for creating manufactured examples and a review of exploratory results that have been utilized to increase compre-hension and direction in this rising field.

11.6.4   Artificial Neural Networks in Urban Runoff Forecast

Rainfall runoff models are highly useful for water resources planning, development and flood mitigations. This exercise is also useful for the prediction of natural calamities like floods and droughts. It also plays a vital role in the design and operation of various components of water resources projects like water supply schemes, barrages, dams, etc. One of the uses of data mining is the extraction of learning from time arrangement. Artificial neural systems have turned out to be suitable in information digging for taking care of the time arrangement. The work displayed by Miguelez et al. (2009) uses artificial neural systems and genetic algorithm with a period in the field of civil engineering where the prescient structure does not take after the exemplary ideal models. In the field of structural building there are a few strategies in view of numerical mathematical statements for displaying the precipitation overflow process, for example, water-driven comparisons (taking into account Saint Vennant mathematical statements) and the unit hydrographs (taking into account exploratory model). The work presented here utilizes artificial neural network and genetic algorithm to hydrology, for the displaying of water stream, created after a downpour occasion, in a given bowl.

11.7 Future: Bio-inspired Computation

The development of complex, highly reliable integrated systems with greatly reduced design cycle times poses a formidable challenge to today's engineers. Traditional engineering design practices are not adequate to meet this new challenge; engineers are increasingly turning to biology for answers. Engineering has always borrowed from nature to provide conceptual examples (for instance, aerodynamics from birds). However, the extraordinary demands placed on today's engineering designs have resulted in the use of biological concepts in a concrete and mathematically defined way. The objective of this chapter is to provide insights on methodologies for the usage in modern design strategies rooted in biological principles. Furthermore, it makes us aware of biologically inspired engineering techniques equipped with multidisciplinary breadth—thus preparing and collaborating technologies from other disciplines.

Bio-inspired computing has tremendous potential in solving engineering issues because of its self adaptability, self-healing and self-nurturing capabilities. Since components of bio-inspired computation tend to cohesively work together in the complex dynamic network, the biologically inspired model aims to achieve the same nature in the networking domain. The networking domain is dynamic and continuously aims in achieving optimum results. Bio-inspired algorithms are based on exploring and mapping the structure, modeling of the system in the networking domain. It has its own merits and demerits. Bio-inspired models aims at developing models which are more flexible, distributed, efficient, and scalable. They show improved performance, scope, and innovation as compared to the optimized systems. Furthermore, the systems so developed are more intelligent, testified, and verifiable under major catastrophic errors. However, since the models are light weight model, scalability, and performance seem to conflict with each other. Component design plays a major role where cooperative versus competitive interactions seem to trade off with each other. Thus, the overall understanding of bio-inspired systems depict that the systems so developed are effective, performance oriented with scalability taken into major account.

From information technology to material engineering, nature and human inspired models have always shown inspiration in form or the other. By deriving analogies from the ecosystem, the expansion of biomimetic material technology has blossomed in reshaping the chemical processing. Furthermore, nature inspired cities and buildings are built for conserving the natural resources. This technology enables less emission of dangerous emissions which increases the life expectancy of humans. Medicine and biology is tightly intervened and bounded with each other and thus have shown great impetus in producing drugs for blindness, bones, cancer, etc. Nonetheless, smart grid technology tightened with bio-inspired models has already conserved energy in some way or the other. Also, mimicking natural systems, such as ants, bees, bats, fishes have inspired many information technology issues.

More aptly, questioning nature in solving human developed models is exactly the way bio-inspired models work. In contrast to the existing models which require regular updating, replacement, nourishing, bio-inspired models maintain themselves and even learn during changing conditions. For instance, ocean urchins make them hone teeth that permit them to bite through rock. On the off chance that we could duplicate the rotating crystalline and natural layers that make up the structure of these teeth, we could construct nanoscale needles that stay pointy even with rehashed use. Maybe in the end we could likewise make bigger scale instruments that do not require manual honing. In the far off future, we may even create machines that repair themselves, like the way skin recuperates over an injury. The future of engineering lies in the development of flexible, self-healing bio-inspired models. As assets become deficient and the vitality emergency comes closer, bio-inspired models gives incalculable chances to enhance the productivity and viability of our advances. On the off chance that we need to move toward a greener future, we ought to look to the green that is as of now around us.

References

Akerlof, G. (2015). Bio-inspired solar energy. http://www.cifar.ca/bio-inspired-solar-energy Accessed on September 15, 2015.

Barreto, M., Jimenez, D., & Satizabal, H. (2006). Andrés Pérez-Uribe, Eduardo Sanchez REDS Institute (http://reds.eivd.ch) *University of Applied Sciences of Western-Switzerland—HEIG-VD The COCH project February 27, 2006.*

Behdad, N., Al-Joumayly, M., & Li, M. (2011). Biologically inspired electrically small antenna arrays with enhanced directional sensitivity. *Antennas and Wireless Propagation Letters, IEEE, 10,* 361–364.

Bizzi, E., Tresch, M. C., Saltiel, P., & d'Avella, A. (2000). New perspectives on spinal motor systems. *Nature Reviews Neuroscience, 1*(2), 101–108.

Burns, L. A., Mouritz, A. P., Pook, D., & Feih, S. (2012). Bio-inspired design of aerospace composite joints for improved damage tolerance. *Composite Structures, 94*(3), 995–1004.

Chan, E. P., Greiner, C., Arzt, E., & Crosby, A. J. (2007). Designing model systems for enhanced adhesion. *MRS Bulletin, 32*(06), 496–503.

Chang, P. C., & Liu, S. C. (2003). Recent research in nondestructive evaluation of civil infrastructures. *Journal of Materials in Civil Engineering, 15*(3), 298–304.

Colgate, J. E., & Lynch, K. M. (2004). Mechanics and control of swimming: A review. *IEEE Journal of Oceanic Engineering, 29*(3), 660–673.

Dawson, T. J. & Taylor, C. R. (1973). *Energetic cost of locomotion in kangaroos.*

Delcomyn, F. (1980). Neural basis of rhythmic behavior in animals. *Science, 210*(4469), 492–498.

Fell, A. (2015). Bio-shock resistant: New center to apply biology to earthquakes. *Civil Engineering, UCDavis.*

Gaylord, N. G., & Eirich, F. R. (1950). Peroxide-catalyzed polymerization of isopropenyl acetate. *Journal of Polymer Science, 5*(6), 743–744.

Grillner, S. (2006). Biological pattern generation: The cellular and computational logic of networks in motion. *Neuron, 52*(5), 751–766.

Honey, K. T. & Pagani, G. A. (2013). *Bio inspired energy.*

Hopkins, J. K., Spranklin, B. W., & Gupta, S. K. (2009). A survey of snake-inspired robot designs. *Bioinspiration & Biomimetics, 4*(2), 021001.

Hu, H., Liu, J., Dukes, I., & Francis, G. (2006). Design of 3D swim patterns for autonomous robotic fish. In *IEEE/RSJ International Conference on Intelligent Robots and Systems* (pp. 2406–2411).

Ishihara, H., Arai, F., & Fukuda, T. (1996). Micro mechatronics and micro actuators. *IEEE/ASME Transactions on Mechatronics, 1*(1), 68–79.

Ishikawa, M., Minami, Y., & Sugie, T. (2010). Development and control experiment of the trident snake robot. *IEEE/ASME Transactions on Mechatronics, 15*(1), 9–16.

Koleva, D. A., Jonkers, H. M., & van Breugel, K. (2013). Bio-inspired control of material properties in civil engineering: Current concept for corrsoion protection via the application of bacteria and bio-polymers. *Ecology & Safety, 7*(1).

Kosa, G., (2015). Bio-inspired sensing for agricultural robots, *BARD Workshop—Innovations in agricultural robotics for precision agriculture.*

Liu, J., & Hu, H. (2010). Biological inspiration: from carangiform fish to multi-joint robotic fish. *Journal of Bionic Engineering, 7*(1), 35–48.

Liu, K., Yao, X., & Jiang, L. (2010). Recent developments in bio-inspired special wettability. *Chemical Society Reviews, 39*(8), 3240–3255.

Loeb, G. E. (2001). Learning from the spinal cord. *The Journal of Physiology, 533*(1), 111–117.

Low, K. H. (2009). Modelling and parametric study of modular undulating fin rays for fish robots. *Mechanism and Machine Theory, 44*(3), 615–632.

Low, K. H., & Willy, A. (2006). Biomimetic motion planning of an undulating robotic fish fin. *Journal of Vibration and Control, 12*(12), 1337–1359.

Miguélez, M., Puertas, J., & Rabuñal, J. R. (2009). Artificial neural networks in urban runoff forecast. In *Bio-inspired systems: computational and ambient intelligence*, (pp. 1192–1199). Springer: Berlin Heidelberg.

MIT (2009). http://ocw.mit.edu/courses/aeronautics-and-astronautics/16–982-bio-inspired-structures-spring-2009/syllabus/. Accessed on September, 24 2015.

Neveln, I. D., Bai, Y., Snyder, J. B., Solberg, J. R., Curet, O. M., Lynch, K. M., & MacIver, M. A. (2013). Biomimetic and bio-inspired robotics in electric fish research. *The Journal of Experimental Biology, 216*(13), 2501–2514.

Noor, A. K., Venneri, S. L., Paul, D. B., & Hopkins, M. A. (2000). Structures technology for future aerospace systems. *Computers & Structures, 74*(5), 507–519.

Odobel, F., Pellegrin, Y., & Warnan, J. (2013). Bio-inspired artificial light-harvesting antennas for enhancement of solar energy capture in dye-sensitized solar cells. *Energy & Environmental Science, 6*(7), 2041–2052.

Otsuka, K., & Wayman, C. M. (1999). *Shape memory materials.* Cambridge University Press.

Renton, W. J. (2001). Aerospace and structures: Where are we headed? *International Journal of Solids and Structures, 38*(19), 3309–3319.

Sanders, J., Scott, E., Weusthuis, R., & Mooibroek, H. (2007). Bio-refinery as the bio-inspired process to bulk chemicals. *Macromolecular Bioscience, 7*(2), 105–117.

Shugen. (2001). Analysis of creeping locomotion of a snake-like robot. *Advanced Robotics, 15*(2), 205–224.

Singh, R., Sharma, R. P., Dutt, S., Kandwal R., & Verma R. (2013). Handbook of agricultural engineering. *Indian Council of Agricultural Research.*

Stein, P. S. (1999). *Neurons, networks, and motor behavior.* MIT Press.

Taylor, C. R., Shkolnik, A. M. I. R. A. M., Dmi'el, R. A. Z. I., Baharav, D., & Borut, A. R. I. E. H. (1974). Running in cheetahs, gazelles, and goats: energy cost and limb configuration. *American Journal of Physiology-Legacy Content, 227*(4), 848–850.

Thompson, V. (2013). Future power grids inspired by the human brain. http://www.livescience.com/27879-bio-inspired-neural-networks-power-grids-brain-awareness-nsf.html. Accessed on September 26, 2015.

Transeth, A. A., Pettersen, K. Y., & Liljebäck, P. (2009). A survey on snake robot modeling and locomotion. *Robotica, 27*(07), 999–1015.

Triantafyllou, M. S., & Triantafyllou, G. S. (1995). An efficient swimming machine. *Scientific American, 272*(3), 64–71.

Ueda, J., Odhner, L., & Asada, H. H. (2007). Broadcast feedback of stochastic cellular actuators inspired by biological muscle control. *The International Journal of Robotics Research, 26*(11–12), 1251–1265.

Wereley, N. (2015). Bio-inspired pneumatic artificial muscles for aerospace and robotic applications.

Whelan, P. J. (1996). Control of locomotion in the decerebrate cat. *Progress in Neurobiology, 49*(5), 481–515.

Wood, R. J. (2008). The first takeoff of a biologically inspired at-scale robotic insect. *IEEE Transactions on Robotics, 24*(2), 341–347.

Yu, J., Ding, R., Yang, Q., Tan, M., Wang, W., & Zhang, J. (2012). On a bio-inspired amphibious robot capable of multimodal motion. *IEEE/ASME Transactions on Mechatronics, 17*(5), 847–856.

Zheng, Y. J., Chen, S. Y., Lin, Y. & Wang, W. L. (2013). Bio-inspired optimization of sustainable energy systems: A review. *Mathematical Problems in Engineering.*

Zhou, C., & Low, K. H. (2012). Design and locomotion control of a biomimetic underwater vehicle with fin propulsion. *IEEE/ASME Transactions on Mechatronics, 17*(1), 25–35.

Index

A

Anomaly detection engine, 144, 146, 149, 156, 157
Ant, 4, 8, 37–39, 74
Ant colony optimization, 4, 37, 38, 74
ART model, 92, 94
Artificial immune systems, 2, 33, 56, 59, 61, 63
Artificial neural network, 46, 79, 81, 83, 89, 93, 94, 191
Attack, 8, 51, 55, 63, 110, 117, 119, 120

B

Back propagation, 137, 138, 83, 88
Bacteria foraging, 43, 47
Bayesian network, 133, 134, 142–144
B-cells, 51, 52, 54, 55, 153
Bees, 1, 4–6, 8, 192
Bee colony optimization, 41, 42
Bio-inspired, 1–4, 7, 74, 105, 117, 119, 143, 161, 177, 180, 182, 183, 185, 191, 192
Bio-inspired aerospace, 182
Bio-inspired agriculture, 180
Bio-inspired civil, 189
Bio-inspired computing, 32, 72
Bio-inspired electrical, 185
Bio-inspired energy, 177, 178, 180, 183
Bio-inspired mechatronics, 187, 188
Bio-inspired systems, 1, 3, 4, 7, 30, 32, 191
Biology, 1–5, 7, 31, 67, 97, 107, 185, 187, 189, 191, 192
Biological evolution, 31
Biologically inspired, 1–3, 6, 7, 8, 48, 144, 160, 183, 185, 191
Bird, 1, 4, 5, 8, 39, 49, 177, 191
Bird colony optimization, 39
Boltzmann learning rule, 85
Bus, 12–14, 19

C

Cells, 28, 30, 34, 51, 52, 55, 57, 58, 152
Clonal selection, 58, 59
Cognition, 8, 30, 112, 113
Communicable diseases, 67, 72
Competitive network, 89
Competitive learning rule, 86
Computer network, 2–4, 6, 14, 51, 89, 187
Crossover, 97, 99, 100, 103
Cuckoo search, 47, 48

D

Data centric faults, 120
Data link layer, 18, 124
Design strategy, 16

E

Epidemic spreading, 2, 4, 6, 8, 67
Error correction rule, 83–85

F

Feed forward networks, 82, 87
Firefly synchronization, 2, 5, 42, 76
Fuzzy logic, 139, 140, 142–144, 180

G

Game theory, 104, 134, 142–144
Genetic algorithms, 97, 99–101, 104–106, 191

H

Hebbian rule, 85
Highly connected nodes, 67, 76
Hopfield network, 90
HOT model, 30
Human immune system, 51, 53, 54, 57, 63

© Springer International Publishing Switzerland 2016
H. Rathore, *Mapping Biological Systems to Network Systems*,
DOI 10.1007/978-3-319-29782-8